Wastewater Sampling for Process and Quality Control

Manual of Practice No. OM-1

Prepared by **Task Force on Wastewater Sampling for Process and Quality Control**

Timothy D. Bradley, *Chair*

Elie Dick	Steven W. Hinton	Roger G. Sigler
Paul L. Eckley	Roger R. Hlavek	David A. Smith
Alan C. Ford	Douglas Lee Miller	W. Griffith Smith
Richard I. Gillette	Katie Mitzner	Frederick D.
Charlene Givens	Thomas O'Grady	Sueverkruepp III
Robert P. Gomperts	Manu A. Patel	Christopher Ward
William L. Goodfellow, Jr.	Charlene Powell	Greg White
John C. Groenewold	Ralph E. Setter	Myra G. Zabec

Under the Direction of the **Operations and Maintenance Subcommittee of the Technical Practice Committee**

1996

Water Environment Federation
601 Wythe Street
Alexandria, VA 22314-1994 USA

Abstract

Sampling is the first step required to effectively monitor and control a wastewater treatment facility. The purpose of this manual is to provide plant operational and management personnel with guidance on contemporary practices for sample collection, sample preservation and handling, data preparation, report generation, and safety.

Chapter 1 describes the purpose and scope of the manual. Chapter 2 presents sampling methods and practices for wastewater, sludge, and gas, and includes a comprehensive listing of recommended process control tests for all common unit processes along with the associated sample type, location, and frequency. Chapter 3 discusses manual and automatic sampling equipment and flow-measuring equipment used in conjunction with sampling equipment. Chapter 4 describes quality assurance and quality control practices regarding wastewater sampling. Chapter 5 describes sample preservation, handling, and shipping. Chapters 6 through 8 deal with data collection, data preparation, and data reporting, from sample labeling and chain-of-custody, to data reduction and quality control, and ultimately to preparation of laboratory, plant, and regulatory reports. Chapter 9 presents safety considerations associated with wastewater sampling.

This manual is intended to be a comprehensive and useful reference source for accurate sampling, which is fundamental to successful process control and achieving maximum plant performance.

Library of Congress Cataloging-in-Publication Data

Wastewater sampling for process and quality control/prepared by Task Force on Wastewater Sampling for Process and Quality Control; under the direction of the Operations and Maintenance Subcommittee of the Technical Practice Committee.

      p. cm. — (Manual of practice. OM; 1)

      Includes bibliographical references and index.

      ISBN 1-57278-037-1 (hardcover)

      1. Sewage—Sampling. I. Water Environment Federation. Task Force on Wastewater Sampling for Process and Quality Control. II. Water Environment Federation. Operations and Maintenance Subcommittee. III. Series: Manual of practice. Operations and maintenance; no. OM-1.

TD735.W27 1996

628.1'61—dc20   96-24226

                                              CIP

ISBN 1-57278-037-1
Printed in the USA **1996**

Water Environment Federation

The Water Environment Federation is a not-for-profit technical and educational organization that was founded in 1928. Its mission is to preserve and enhance the global water environment. Federation members are more than 42,000 water quality specialists from around the world, including environmental, civil and chemical engineers, biologists, chemists, government officials, treatment plant managers and operators, laboratory technicians, college professors, researchers, students, and equipment manufacturers and distributors.

For information on membership, publications, and conferences, contact
Water Environment Federation
601 Wythe Street
Alexandria, VA 22314–1994 USA
(703) 684–2400

Manuals of Practice for Water Pollution Control

The Water Environment Federation (WEF) Technical Practice Committee (formerly the Committee on Sewage and Industrial Wastes Practice of the Federation of Sewage and Industrial Wastes Associations) was created by the Federation Board of Control on October 11, 1941. The primary function of the committee is to originate and produce, through appropriate subcommittees, special publications dealing with technical aspects of the broad interests of the Federation. These manuals are intended to provide background information through a review of technical practices and detailed procedures that research and experience have shown to be functional and practical.

IMPORTANT NOTICE

The contents of this publication are for general information only and are not intended to be a standard of the Water Environment Federation (WEF).

No reference made in this publication to any specific method, product, process, or service constitutes or implies an endorsement, recommendation, or warranty thereof by the Federation.

The Federation makes no representation or warranty of any kind, whether expressed or implied, concerning the accuracy, product, or process discussed in this publication and assumes no liability.

Anyone using this information assumes all liability arising from such use, including but not limited to infringement of any patent or patents.

Water Environment Federation Technical Practice Committee Control Group

L.J. Glueckstein, *Chair*
T. Popowchak, *Vice-Chair*

G.T. Daigger
P.T. Karney
T.L. Krause
G. Neserke
J. Semon
R. Zimmer

Authorized for Publication by the Board of Control
Water Environment Federation
Quincalee Brown, *Executive Director*

Preface

This manual of practice presents contemporary considerations for wastewater sampling. The manual, which replaces the previous wastewater sampling operations manual published in 1980 by the Water Environment Federation (then Water Pollution Control Federation), covers sampling methods and practices, sampling equipment, sample preservation and handling, data collection and preparation, report generation, safety, and quality assurance and quality control.

The manual is intended for the practicing wastewater professional. Accordingly, it focuses on current practices related to all aspects of sampling and is intended for those individuals with a general knowledge of the basic principles of sampling. Considerations discussed are intended to span the full range of activities associated with sampling, from designing and implementing a sampling program through all subsequent steps required to ensure that the sample data are accurate and properly documented and reported.

The contents of this manual represent the collective background and experience of numerous professionals active in wastewater sampling. Material presented is intended to complement other Water Environment Federation manuals of practice and published plant operations and process control guidelines and textbooks.

This manual was produced under the direction of Timothy D. Bradley, *Task Force Chair*. The principal authors are

Timothy D. Bradley	Robert P. Gomperts	David A. Smith
Brett DeHollander	Douglas Lee Miller	Frederick D.
Paul L. Eckley	Thomas O'Grady	Sueverkruepp III
Richard I. Gillette	Charlene Powell	Myra G. Zabec

In addition to Task Force and Technical Practice Committee Control Group members, additional information and review were provided by

George Cartledge
Murali Kalavapudi

Authors' and reviewers' efforts were supported by the following organizations:

Advanced Environmental Services, Niagara Falls, New York
Calgon Corporation, Pittsburgh, Pennsylvania
Chester Environmental, Pittsburgh, Pennsylvania
City of Baltimore, Maryland
City of Edmonton, Alberta, Canada
City of Salem, Oregon
EA Engineering, Science, and Technology, Sparks, Maryland

Earth Tech, Beckley, West Virginia
Eastman Kodak Company, Rochester, New York
Enviro/Sci Corporation, Radnor, Pennsylvania
Givens & Associates, Cumberland, Indiana
HYDROSAMPLE, Division of Zecco, Inc., Northboro, Massachusetts
Isco, Inc., Lincoln, Nebraska
Kraft General Foods, Northfield, Illinois
Malcolm Pirnie, Inc., White Plains/Syracuse, New York
Metcalf & Eddy, Inc., Somerville, New Jersey
Operations Management International, Inc., South Portland, Maine
Stony Brook Regional Sewerage Authority, Princeton, New Jersey
Tufts University, Department of Civil Engineering, Carlisle, Massachusetts
Woodard & Curran, Inc., Portland, Maine

Federation technical staff project management was provided by Bill Nivens; technical editorial assistance was provided by Matthew Hauber.

Contents

List of Tables

List of Figures

Chapter 1
Introduction

Sampling is the first step required to monitor and control a wastewater treatment facility effectively. Reliably achieving maximum performance requires implementation of and adherence to proper sampling procedures. This manual provides plant operational and management personnel with guidance on contemporary practices for designing and implementing an effective sampling program, including sample collection, sample preservation and handling, data preparation, report generation, safety, and quality assurance and quality control.

Representative sampling is fundamental to the overall success of any attempt to monitor plant performance or control its treatment processes; however, its importance is often overlooked. Like its predecessor, this updated manual emphasizes important fundamentals and provides the reader with a useful reference source. It is not necessary to read the manual from cover to cover, as each chapter addresses a stand-alone component of the sampling process.

Chapter 2 presents methods and practices for obtaining representative samples. Sample types and volumes are discussed, and recommendations are given for sampling location and frequency. A comprehensive list of recommended process control tests for all common unit processes is also included as well as plant design considerations for sampling. This chapter deals with the initial steps in the design of a comprehensive sampling program.

Chapter 3 discusses manual and automatic sampling equipment in addition to flow-measuring equipment used in conjunction with sampling equipment. Because of advances in automatic sampling equipment since the initial publication of this manual, considerable detail is included on this topic. Likewise, the flow measurement section has been updated to include relatively new flow meters such as velocity-area and the Doppler type. This chapter also addresses various on-line sampling equipment in common use today.

Chapter 4 deals with quality assurance and quality control in sampling. Topics presented in this chapter relate to sampling precision and accuracy and include sources of error, sample duplicates, trip blanks, field blanks, blind standards, field split samples, field check standards, equipment calibration samples, and documentation and recordkeeping.

Chapter 5 describes sample preservation techniques to minimize errors associated with biological, chemical, and physical changes in the collected sample. Because samples are frequently sent to outside laboratories for analysis, procedures for proper sample handling, packaging, and shipping are also included, along with a discussion of pertinent transportation regulations.

Chapter 6 presents data collection techniques that are essential to ensuring that properly labeled samples are delivered to the laboratory for analysis. A detailed discussion of sample labeling guidelines and procedures is included along with a discussion of automated data-recording systems. The chapter concludes with a discussion of internal and external audit considerations.

Chapter 7 provides useful information about data gathered from wastewater sampling and describes how to perform simple statistical calculations to determine the quality of data. This procedure is important to ensure that meaningful data are used for process control decisions and that data presented on National Pollutant Discharge Elimination System forms have been properly recorded and analyzed. Along with data reduction and statistical evaluation, quality control and graphing techniques are presented. The chapter also describes computer software available for performing statistical analyses.

Chapter 8 addresses methods of compiling and handling data. Because most data are ultimately presented in laboratory, in-plant, or regulatory reports, a detailed discussion of report preparation is included. The use of computer-based systems to store and analyze data and generate reports is becoming commonplace in larger plants. Therefore, a discussion of laboratory information management systems is also included.

Finally, Chapter 9 presents safety practices pertinent to sampling. Wastewater sampling can expose the individual collecting the samples to a potentially dangerous environment, including waterborne diseases. Therefore, implementation of and adherence to proper safety procedures are of utmost importance to any sampling program. Topics presented in this chapter include training, housekeeping, work procedures, accidents/incidents, and emergencies.

Chapter 2
Sampling Methods and Practice

Sampling is performed for a wide variety of purposes in water, wastewater, and industrial waste treatment plants. Some of the more common purposes for sampling include regulatory reporting requirements, plant process control, checks on industrial processes, toxicity evaluations, and dispersion studies. Sampling programs at wastewater treatment plants vary with the size of the facility, the available sampling and laboratory personnel, the purpose of sampling, regulatory reporting requirements, and the complexity of the treatment processes to be sampled. An accurate and meaningful sampling program is essential to produce "defensible" pollutant analyses and is also important for process control in a well-operated treatment facility.

Not only are process sampling and testing important, but the proper understanding and use of the resulting data are just as essential. The major parts of a good sampling program include defining sampling goals and requirements, preparing a written sampling plan, conducting representative sampling, preserving samples properly, and performing meaningful analyses. A good sampling program will help in making process adjustments and meeting regulatory requirements.

Some treatment plants only perform the sampling and analyses required to fulfill the facilities' regulatory discharge permits or requirements. However, this is usually inadequate for good plant operation. Proper operation of a treatment plant requires knowing what is occurring with the various processes in the plant, not just testing the quality of the effluent. Knowledgeable process adjustments can only be made with information on efficiencies and loadings of various plant processes that has been gained through a good sampling program.

*F*UNCTION OF SAMPLING

The purposes and functions associated with sampling include compliance with regulatory requirements, process monitoring and control, historical data collection, collection system studies, and personnel safety programs. All of these purposes of sampling must be considered when establishing a meaningful sampling program.

REGULATORY REQUIREMENTS. Treatment plants that operate under regulatory permits are usually required to perform specific sampling and analyses on a regular basis. Permits usually specify the sampling location, type (grab or composite), and frequency; required parameters; methods of analysis; and the required frequency in reporting the analyses to the regulatory agency. National Pollutant Discharge Elimination System (NPDES) permits also specify that sampling and analysis must comply with either the U.S. Environmental Protection Agency's (U.S. EPA's) manual on *Methods for Chemical Analysis of Water and Wastes* (Kopp, 1983) or the American Public Health Association's *Standard Methods for the Examination of Water and Wastewater* (1995) as specified in the Code of Federal Regulations, 40 CFR Part 136. Typically, the sampling and analyses specified in an NPDES permit concern only the effluent from a treatment plant.

PROCESS MONITORING. The sampling and analyses required to fulfill most permits will normally reveal how well processes at a treatment plant are performing as a whole. However, to properly operate a treatment plant and control each unit process, many more testing parameters must be developed. An effective sampling program provides total system information on the loading (influent characteristics), performance (effluent quality), and intermediate conditions of each unit process within a treatment plant. The more information known about the process performance and waste stream characteristics as it progresses through the treatment plant, the better informed an operator will be to make proper operational adjustments. The operational staff must be knowledgeable enough to use the sampling and test results to make the proper operational adjustments. Too often, personnel perform tests but do not use the information to make informed decisions.

Many of these process control measurements can be made in the field for more accurate and faster results. For example, using a portable dissolved oxygen meter to measure dissolved oxygen in an aeration tank and measuring a sludge blanket level directly in a clarifier are typical field process control tests.

TREND ANALYSES. A database of historical sampling and analyses is invaluable to operators and engineers. To the operator, a historical database can show seasonal variations at the treatment plant, pointing out past abnormal conditions that can help in preparing for treatment process adjustments. A historical database can also point to corrective actions that have been used in the past for recurring abnormal conditions: it can reveal which corrective actions have worked and which have not worked. Historical records will show trends in plant loading and performance that can be used to predict when a plant expansion or upgrade is needed. Also, the database will supply information useful in designing a plant upgrade or expansion.

The historical database is simply the retained records of past sampling and testing. Of course, if process control tests are not performed or rarely performed, little information will be available for the historical database. It should be noted, however, that any historical data must be considered with regard to possible changes in treatment plant influent characteristics or in-plant system changes that may permanently alter the normal conditions. Also, historical database comparison should not be used if vastly different methods or detection limits have been used.

COLLECTION SYSTEM STUDIES. Collection system sampling generally focuses on contaminant discharges to receiving waters through combined sewer overflows and outfalls. Other collection system sampling concentrates on the isolation of high parameter loadings, possible illegal connections, or cross-connections. Sampling and subsequent sample analysis in isolated areas of a collection system can help reveal or even pinpoint the source of high loadings.

PERSONNEL SAFETY. Samples must be taken of atmospheric conditions before entering and during occupancy of a confined space or areas where harmful gases are likely to exist. These gaseous samples are often taken with portable meters that analyze a gas sample and sound an alarm when dangerous conditions exist. Typical analyses include those for oxygen, hydrogen sulfide, carbon monoxide, and explosive conditions. The samples to be taken depend on the type of area entered. Consult a safety manual or safety expert on confined space entry procedures before making any confined space entries or any entries into areas that could possibly collect harmful gases.

DESIGN OF A SAMPLING PROGRAM

A written sampling plan should be prepared before starting any program. A sampling plan improves the quality of the program by clearly outlining all sampling factors. Sampling plans should include the following:

- Why are samples collected? What is the purpose of sampling? Is it for regulatory agency compliance, routine process control information, determining user fees, or testing for pollutants? The reason for sampling and frequency of sampling should be outlined completely and described in detail.
- What parameters will be analyzed? The types of pollutants to be analyzed will determine the volume of sample required. The specific parameters will also determine how the sample must be preserved, how long the sample can be held before analysis, and in what type of container it must be stored.
- What are the sample site conditions? The sample site conditions influence the exact sampling location as well as what sampling equipment will be used. If the site is not easily accessible, the equipment may have to be set up or programmed before it is placed.
- What type of sample will be collected? Will a discrete grab sample or composite sample be required? This decision is often dictated by regulatory requirements or a requirement of the analysis to be performed (for example, chlorine residual or dissolved oxygen).
- What equipment will be used? Will the sample be taken automatically or manually? In what type of container will the sample be stored? Is a flow-measuring device required? What will trigger the sampling event? These are all important questions to ask before initiating a sampling program. The frequency of sampling will usually determine the sampling equipment to be used. For example, if sampling is infrequent, manual sampling or the use of temporary equipment may be the best choice. Other items such as gloves, protective equipment or clothes, sample log or clipboard, chain-of-custody records, and special tools must also be determined.

TYPES OF SAMPLES

To determine what type of sample to collect, the operator must decide what analyses are to be performed and what information is needed. There are two basic types of wastewater samples: grab or composite. Both types of samples can be collected either manually or through the use of an automatic sampling device. Composite samples can be collected in proportion to flow or based on a set time interval.

In general, the selection of the sample type to be collected is governed by the information needed, the unit process being sampled, the analyses to be performed, regulatory permit stipulations, whether the flow is continuous or intermittent, and how flow characteristics vary over time. Sampling-site accessibility and safety must also be considered when choosing one particular site over another.

GRAB SAMPLES. A grab sample is one that is taken to represent one moment in time and is not mixed with any other samples. A grab sample is sometimes called an *individual* or *discrete* sample and will only represent sample conditions at the exact moment it is collected.

Grab samples are also often useful under certain other situations for which composite samples would not be adequate. Examples of these situations include the following:

- The characteristics of "slug" discharges must be determined by grab samples to help identify the source and assess the potential effects on the treatment processes. These discharges are often noticed visually by a plant operator performing routine duties, and the duration typically is unknown.
- Numerous grab samples are used to study variations and extremes in a waste stream over a period of time. Composite samples do not reveal waste variations over time because of the nature of the samples. Composites tend to average both short-duration, high-strength discharges and long-duration, low-strength discharges. The significance of this depends on the volume of flow at the time of collection. Composite samples tend to dilute short, high- or low-strength discharges that could affect a treatment plant's performance but go unnoticed. Samplers that can take both composite and sequential discrete samples can be advantageous in periodically testing for waste variations over a specific time frame.
- Grab samples can be used if the flow to be sampled occurs intermittently for short durations.
- Grab samples can be used if the flow composition to be sampled is reasonably constant. Of course, this assumption must be verified with

---

multiple samples over an adequate time period to determine if there really are variations in flow composition.

- Grab samples should be used if the constituents to be analyzed are unstable or cannot be preserved and, therefore, must be analyzed immediately or stored under special conditions. Examples of these parameters include oil and grease, pH, chlorine residual, dissolved oxygen, bacteriological tests, purgeable organic compounds, and phenols.

COMPOSITE SAMPLES. A composite sample is prepared by combining a series of grab samples over known time or flow intervals. A composite sample shows the average composition of a flow stream over a set time or flow period if the sample is collected proportional to flow. These samples can be collected manually and mixed together or can be collected by automatic sampling equipment. The samples taken by automatic composite sampling equipment are composited as they are collected into one large receptacle. See Chapter 3 for more information on automatic sampling equipment.

At most wastewater treatment plants, composite samples are required under regulatory permit requirements for most constituents that do not require immediate analyses. Typical composite sampling is required for parameters such as biochemical oxygen demand (BOD), suspended solids, ammonia nitrogen, and total phosphorus.

Results from analyses used to calculate plant and process loadings (such as organic loading or food-to-microorganism ratios) should always be made from composite samples. This practice is important in ensuring that data obtained from a slug or spike flow, using a single grab sample, do not bias the information or provide misleading data.

Two different types of composite samples are generally used: fixed volume or flow proportional.

Fixed-Volume Composite Samples. The fixed-volume composite, also called a *time composite*, is the simpler type of composite sample. In fixed-volume composite sampling, a series of individual grab samples, all having the same volume, are collected at equally spaced time periods. Fixed-volume composite samples will only give an accurate representation of average flow characteristics if flow does not vary during the sampling period. This is not often the case in typical treatment plants, even when systems are equipped with flow-equalization tanks to dampen flow variations.

The fixed-volume composite sample is more appropriate for sampling activated-sludge aeration basins, sludge solids in digesters, constant-flow streams, and sludge cakes from dewatering equipment.

The total volume of the composite sample required depends on the types of analyses that must be performed on the sample. The number of individual grab samples required to make up the composite sample depends on the time frame of the sampling event and other factors such as regulatory requirements and degree of accuracy. U.S. EPA allows time-proportional sampling and in its standards requires 15-minute intervals (96 samples/day). Generally speaking, the more individual samples collected, the better the composite will represent the flow stream. For example, 24 samples of 500 mL each collected to form a 12-L composite will better represent the flow stream than 12 samples of 1 L each.

To calculate the volume of each individual sample that must be collected and combined into one composite sample, the time interval and total composite sample size required must first be determined. For example, if a 1-L composite sample must be collected over a 24-hour period, with sampling intervals every 2 hours, the calculation is as follows:

Number of samples collected =

$$\frac{\text{Total hours}}{\text{Sample frequency}} = \frac{24}{2} = 12 \text{ samples} \tag{2.1}$$

Volume of each sample =

$$\frac{\text{Total sample volume}}{\text{Number of samples}} = \frac{1\ 000 \text{ mL}}{12 \text{ samples}} = 84 \text{ mL/sample} \tag{2.2}$$

Allowing for surplus sample, the volume should be rounded up to 100 mL. When rounding up, the container size and accuracy of collection volume must be considered.

When individual grab samples are combined as a composite, they must be transferred quickly, first from the collection point to the sample-measuring device and then to the composite container, while continuously being mixed. Agitating the sample in this manner prevents settling and reduces sampling error. This process is best accomplished using automatic composite-sampling equipment, which eliminates the measure-transfer step.

Flow-Proportional Composite Samples. In flow-proportional sampling, the sample volume collected varies based on the flow rate of the waste stream being sampled. Either the volume of each individual grab sample or the sample frequency is varied in direct proportion to flow rate. To be correct, a flow-proportional composite sample must be based on accurate measurements of the waste stream flow rate.

A flow-proportional composite sample is more representative of the waste stream than a fixed-volume sample because it takes into account variations in wastewater characteristics that result from fluctuations in flow. Many NPDES permits require that flow-proportional samples of the plant effluent be collected. Typical parameters that are often analyzed in flow-proportional composite samples include suspended solids, BOD, ammonia nitrogen, and total phosphorus.

VARIABLE-VOLUME TECHNIQUE. If manual sampling is used to create a flow-proportional composite sample, the variable-volume technique is usually the easier and more practical way to collect the sample. With this technique, the volume of each sample collected is based on the flow rate of the waste stream at the instant the sample is collected. An example of one procedure used to manually collect a variable-volume composite sample is given below.

Steps to Composite Sampling Using Variable Volumes.

1. Determine the minimum composite sample volume, in millilitres, needed to perform the desired analyses. Refer to *Standard Methods for the Examination of Water and Wastewater* (APHA, 1995) to determine the required sample volumes for each analysis. An additional volume of sample should be collected to allow for duplicate quality control analysis or spillage of the sample.

2. Divide the composite sample volume required, in millilitres, by the expected average flow rate (in gallons per day [gpd], thousand gallons per day, or million gallons per day [mgd]), for the period to be sampled. The number calculated equals the sample volume per unit volume of flow (in millilitres per million gallons per day [mL/mgd]).

3. Determine the number of samples required during the sampling period by dividing the composite sampling period, in hours, by the interval of each sampling event, in minutes or hours.

4. Divide the resulting value from step 2 by the number of samples to be collected during the sampling period calculated in step 3. The result represents the volume of each sample (mL) to be collected per unit of flow (gpd, thousand gpd, or mgd) at each sampling event. Round the answer calculated up to the nearest whole number.

5. To determine the amount of sample to collect at each sampling interval, multiply the flow (gpd, thousand gpd, or mgd) at the time of sampling by the value calculated in step 4.

6. Develop a chart or table showing the volume of sample to be collected at each of several different flow rates that are possible during the sample period. The table should be developed by multiplying each flow rate by the number calculated in step 5. Preparing a table or chart will make it easier to determine quickly the sample volume for each sampling event.

Example. An example using the above procedure is presented below. For the example, an 8-hour, flow-proportional composite sample is to be taken at a sampling interval of once every 15 minutes. The parameters to be analyzed are BOD, total suspended solids (TSS), and ammonia-nitrogen (NH_3-N). The daily average flow at the treatment plant is 5 677 m^3/d (1.5 mgd), and the minimum measurable reading is 378.5 m^3/d (0.1 mgd).

1. The final minimum sample volume required is calculated based on the volume required to perform each laboratory test:

BOD	TSS	NH_3-N
1 000 mL	+ 100 mL	+ 400 mL = 1 500 mL

2. Divide 1 500 mL, as found in step 1, by 1.5 mgd, which equals 1 000 mL/mgd.

3. Divide the number of samples required in an 8-hour (480-minute) sampling period by a sampling rate of once every 15 minutes: 480 minutes ÷ 15 minutes per sample = 32 samples. A total of 32 samples must be taken during the 8-hour period.

4. Divide the 1 000 mL volume determined in step 2 by the 32 samples calculated in step 3: this yields a value of 31 mL/mgd/sample. Round up to obtain a convenient sample quantity and allow for a safety factor in case the sample is spilled or misused. This yields 50 mL/mgd/sample.

5. To determine the amount of sample to be collected, multiply the flow at the time of sampling by 50 mL, calculated in step 4. For example, if the flow is 1.3 mgd, the sample collected should be 1.3 mgd × 50 mL/mgd = 65 mL.

6. Develop a sample proportioning chart that spans the diurnal, or day and night, range of flow variations through the plant. An example of a flow-proportional chart used for composite sampling developed from the example is given in Table 2.1.

Table 2.1   Flow-proportional sampling chart.

Flow rate, mgd[a]	Sample volume, mL	Flow rate, mgd	Sample volume, mL
0.1	5	1.6	80
0.2	10	1.7	85
0.3	15	1.8	90
0.4	20	1.9	95
0.5	25	2.0	100
0.6	30	2.1	105
0.7	35	2.2	110
0.8	40	2.3	115
0.9	45	2.4	120
1.0	50	2.5	125
1.1	55	2.6	130
1.2	60	2.7	135
1.3	65	2.8	140
1.4	70	2.9	145
1.5	75	3.0	150

[a] $mgd \times (3.785 \times 10^3) = m^3/d$.

If the plant flow rate varies significantly on a seasonal or other basis, it may be necessary to recalculate the steps above. In this way, sample volume-to-flow factors can be developed for each season.

If the flow is not known or is intermittent, as is commonly the case with industrial waste streams, equal volumes of samples can be collected and stored. After the hourly flow rates are obtained, a composite sample can be prepared by mixing the proportion of each sample that corresponds to the flow at the time of collection. For example, 1-L samples can be collected at hourly increments, and the flow can be recorded each time a sample is taken. After the sampling period has ended, the samples can be proportioned according to the recorded flow each time a litre sample was taken. This procedure requires several clean containers and a large storage area. Also, this procedure must be carefully performed to make sure all of the samples are well mixed before composite sampling. This method is better suited to use with automatic sampling equipment, which can collect hourly discrete samples in separate bottles for later composite sampling.

VARIABLE-FREQUENCY TECHNIQUE. In variable-frequency, flow-proportional composite sampling, the volume of sample collected stays constant, but the time interval between samples or frequency in which the samples are collected varies. The sample intervals are proportional to the measured flows. Variable-frequency, flow-proportional composite samples are rarely used with manual sampling but often used with automatic sampling equipment. To collect these types of samples, an automatic sampler is coupled to a flow meter that collects each sample after a preprogrammed amount of wastewater flows past the sampling point. To determine the frequency of each individual sample, the average plant flow, individual sample size, and total composite sample required must be known or accurately estimated. A procedure for determining the wastewater flow volume for each sample interval is given below.

1. Determine the final composite sample volume required: this is determined based on the required volume needed to test all of the parameters to be analyzed.
2. Divide the total composite sample volume by a realistic and convenient volume selected for each sample: this value represents the number of samples to be collected each day.
3. Divide the average daily flow rate of the waste stream by the number of samples to be collected. This number should be adjusted upwards to greater than 100 to be statistically correct for a 24-hour period, or it should be adjusted to whatever is statistically correct for a shorter time period, usually greater than 20 (see Chapter 7 for a discussion on sampling and statistics). This number equals the volume of wastewater that must pass through the sampling point measuring device before a sample is collected. It can be seen from this

procedure that the faster a specific quantity of wastewater passes the sampling point, the more frequently a set-volume sample must be collected.

REPRESENTATIVE SAMPLING

Information obtained through sampling and subsequent analyses is only as good as the techniques applied. If a sample is not collected or preserved properly, it will yield erroneous information. Using process data developed from improper samples could possibly result in incorrect process control decisions, ultimately affecting the performance of the treatment plant. For example, suppose a sample taken from an aeration tank in an activated-sludge process is sampled at an incorrect location, unrepresentative of the actual process, and tested for suspended solids. The operators' analyses indicate a higher than normal suspended solids concentration. These test results are then used to determine the amount of waste sludge to be removed from the system. Wasting too much sludge based on erroneous samples will ultimately have a severe negative effect on the way the process performs. The sample must represent the actual process or true flow stream as closely as possible to provide correct information.

The primary goal of proper sampling is to ensure that the sample collected represents the flow stream being analyzed. To collect representative samples, both proper sample site selection and sampling technique are critical. Several guidelines for representative sampling are listed here and should always be followed.

- Collect samples for nonvolatile constituents at points where the sample stream or tank is well mixed. However, samples should not be collected at the point of maximum turbulence or at the edge of tanks or channels because neither of these areas yields representative samples. Typically, samples in channels should be collected at a point one-third the liquid depth from the channel bottom and midway across the channel between the point of maximum turbulence and the edge.
- Samples collected for the analysis of volatile organic compounds (VOC) should be taken from areas of low turbulence to reduce the amount of entrapped air in the sample. Volatile organic compounds could be driven off to the atmosphere (as outgas) in turbulent sections of the flow stream. Also, if air is entrained in the water, the samples collected cannot be used for VOC analysis because bubbles will develop in the sample container. The "no head space" requirement for VOC sample containers, meaning no air space in the container, is important to ensure that all VOCs are kept in solution for proper analysis.

- Avoid taking samples at points where solids settling occurs or floating debris is present. These situations occur normally in quiescent areas, where the velocity of the flow has decreased.
- Avoid sampling nonrepresentative deposits or solids accumulated on channel or tank walls.
- Collect wastewater influent samples at a point upstream of any location at which recycled process streams such as digester supernatant, trickling filter recycle, or waste activated-sludge return to the main process flow stream.
- After selecting a representative sampling location, sample the waste stream consistently at this location. By maintaining this consistency, variations in sample results cannot be attributed to changes in location. The test data can be compared confidently from day to day. The sample site should be permanently marked with paint or a sign to ensure that everyone samples at the same location.
- Accessibility and safety are also important factors when selecting a sampling site. Do not choose a sample site that is difficult to get to or can result in falls and injuries.
- Flush or purge sample lines for an adequate time period before taking the sample to ensure that material left in the line from the prior sampling event is not incorporated into the sample. Also, replace sample lines regularly to avoid the possibility of sediment buildup, which could cause erroneous results.
- Where samples are to be collected from flowing pipes, keep the sample lines as short as possible and with a minimum number of bends. Excessively long sample lines pose the risks of inadequate flushing before sampling and alterations in the sample resulting from chemical and biological activity within the sample pipe.
- Clearly identify and mark sample containers for each location. This precaution will reduce the possibility of confusing samples from one location with samples from another or mistaking the origin of a sample. Each sample container should be clearly labeled with the date, time, sample location, parameters to be analyzed, and person who collected the sample. Refer to Chapter 6 for additional information on sample labeling.
- Use a different sampling device at each sampling location. Rinse the sampling and measuring devices with fresh sample material before transferring the actual sample to its container, unless preservatives have been added previously to the sample bottle. Sampling devices should be regularly cleaned and replaced. Following these steps will reduce the risk of contaminating samples collected from different areas.
- To ensure that the sample is representative, prevent settling by keeping samples thoroughly mixed throughout the collection and measurement procedure.

- After collection, samples must be properly preserved and stored. Depending on what constituents are to be analyzed, composite samples may need to be kept refrigerated during collection. Refrigerating and preserving the sample helps ensure that its composition does not change before testing. Refrigeration at 0 to 4°C is prescribed by U.S. EPA for permit-required samples and is usually appropriate for samples collected for other purposes. Refer to Chapter 5 for additional information on sample preservation.

WASTEWATER SAMPLING IN TREATMENT PLANTS. Automatic samplers are ideal for sampling the influent and effluent waste streams of a treatment plant. The composite samples typically required for these areas can be collected more easily and more accurately using automatic sampling equipment than by manually collecting the samples.

Grab sampling of wastewater for process control information in a treatment plant often involves the use of sampling equipment such as dippers and weighted bottles. The type of sampling device needed is determined by the wastewater conduit or containment to be sampled, the distance above the sampling point from an accessible area, and the volume of sample required.

Other measuring devices that can be used are those that allow *in situ*, or in-place, measurement of process parameters rather than requiring that samples be collected at the flow stream and taken to a laboratory for analyses. Examples of such devices are dissolved oxygen (DO) or pH meters equipped with field probes or permanently mounted and outfitted for continuous monitoring. Continuous-monitoring, or on-line, analyzers include those for measuring DO, pH, residual chlorine, conductivity, temperature, total organic carbon, and oxidation reduction potential, to name a few. These may also be linked to a controller for automatic adjustment of chemical feed rates or equipment operation. On-line analyzers are becoming increasingly common as technological advances make them more economically feasible and widely available.

When either portable *in situ* meters or on-line analyzers are used, the measuring instrument must be located in a well-mixed, representative area of the waste stream. Both portable meters and on-line equipment must be calibrated and cleaned as recommended by the equipment manufacturer. When positioning an on-line probe or measuring device, follow the same guidelines previously mentioned for collecting representative samples. Refer to Chapter 3 for additional information on sampling equipment.

WASTEWATER SAMPLING IN COLLECTION SYSTEMS. Samples collected in sanitary sewers must be taken in areas of turbulence where solids settling is not occurring. Samples should be taken a sufficient distance from bends in the sewer line to prevent the hydraulics of the wastewater flow from affecting the samples. Generally, the distance required from bends is two pipe diameters upstream or ten pipe diameters downstream of a bend or anything that causes interference. Safety is of utmost importance when sampling in manholes or other confined spaces, and confined space entry procedures must be followed when it is necessary to enter a manhole or sewer.

SLUDGE SAMPLING IN WASTEWATER TREATMENT PLANTS.
Sludge samples are taken in wastewater treatment plants to obtain information on sludge-handling facilities for process control and regulatory reporting requirements. Often, process control samples are overlooked because the solids are not part of the liquid processes. However, proper operation of digesters, thickeners, and dewatering equipment necessitates process sampling and analysis.

Sampling and testing requirements for process control include sampling of primary and secondary sludges, digester feed sludge, digested sludge, dewatered sludge, and various solids-processing side streams such as supernate and filtrate. These samples are often taken from a sludge-transfer line or from the discharge from pumps. As with liquid stream samples, sludge must be well mixed, and it should be taken consistently from the same location to allow comparison of testing results.

Sample bottles should have wide mouths to accommodate the sampling device or large particles of solids. A space for gas collection should be left at the top of the bottle, and the cap should be left loose for gas to escape, if protocol permits (not in VOC analysis). Be aware that bacterial activity in sludges can cause gas production, which results in pressurized sample containers. Care must be taken so the containers will not explode if filled completely.

Sampling and analysis for regulatory reporting requirements are typically performed on the dewatered sludge before landfilling, land application, or incineration. This dewatered sludge cake is the product remaining after belt filter press, vacuum filter, centrifuge, and drying or composting operations. Requirements for sampling are given in the Code of Federal Regulations under the section *Use or Disposal of Sewage Sludge*, 40 CFR Part 503. These regulations require that a valid sample of wastewater sludge be taken from the correct location, represent the entire amount of sludge, and be handled properly from the time of collection through analysis.

The techniques for sampling sludge vary depending on whether the sludge is flowing or solidified in a pile or bin. Sludge that is flowing or moving should be sampled at equal intervals during the length of time the sludge process unit operates in a day. Sludge from a pile or a bin must be collected in the appropriate number of sample aliquots (sample portions) from various points in the pile. The statistical procedure for determining the required sampling is described in the U.S. EPA manual, *POTW Sludge Sampling and Analysis Guidance Document* (U.S. EPA, 1989). At a minimum, full-core samples should be taken from at least four points in the pile or bin.

GAS SAMPLING. The gas from anaerobic digesters is commonly sampled for its composition. The typical gas composition from an anaerobic digester includes methane (CH_4) and carbon dioxide (CO_2). The air in wastewater collection systems, covered tanks, and uncovered empty tanks must be monitored for safety purposes before entry and during occupancy. Common gas composition tests include those for oxygen content, lower explosion limit, and methane and hydrogen sulfide concentrations. These tests are often conducted using portable or on-line gas monitors that take the samples using a small gas pump. Regular calibration and maintenance of air-monitoring equipment is essential for accurate and reliable measurements. Before entering a confined space, such as a closed tank or manhole, the air should be tested at several levels and locations to ensure that the measurements made represent the true conditions within the confined space. Gas monitoring must be performed continuously while a confined space is occupied.

Monitoring of air around treatment plants has become increasingly important because of treatment plant odor concerns and complaints. Volatile chemicals generated at wastewater treatment facilities are ever-increasing concerns too. Sampling and analysis of gaseous air constituents can involve the simple use of a sniffer-type sample tube. This instrument draws an air sample through a tube filled with chemical reagents that change color in proportion to the concentration of a specific compound in the air. A more involved method of air sampling requires collecting a sample with an air pump and capture bag system, which is then tested either using gas chromatography or mass spectrometry systems or by the evaluation of an odor panel.

RECOMMENDED TREATMENT PLANT SAMPLING

The factors that dictate where and how many samples should be collected include regulatory requirements, the size of the facility, plant performance, the intended use of the sampling data, and the plant laboratory capabilities. Of course, the type and quantity of samples required on a routine basis will influence what analyses the plant laboratory should be capable of performing.

The number of samples that must be taken for regulatory reporting requirements is outlined in the regulatory discharge permit. However, the specific parameter limitations listed in the permit will also influence the type and number of process control tests that need to be performed. For example, if the requirements include removal of ammonia–nitrogen, nitrogen, or phosphorus, more analyses are required at different stages of treatment to ensure that the parameter is being properly reduced. In the same manner, if the treatment plant must maintain a high degree of contaminant removal, more process

---

control sampling and testing is required on a regular basis to maintain tighter or stricter control of the plant processes.

The size of the facility and availability of laboratory staff are also factors that influence the sampling program procedures: generally speaking, the larger and more sophisticated the treatment facility, the greater the quantity of sampling typically required and, therefore, the greater the number of laboratory staff needed to perform these duties. *Remember, all samples taken according to proper protocol must be reported.*

How well a plant operates without constant operational adjustment is also a factor in determining the quantity of sampling and analyses needed. A treatment plant that operates well on a continuous basis and with little operator input will require less process sampling and testing than a plant in which operational adjustments must be made constantly or one that does not historically perform well.

SUGGESTED SAMPLING LOCATIONS AND FREQUENCIES IN WASTEWATER TREATMENT PLANTS. Suggested sampling locations and frequencies for common wastewater treatment unit processes are listed in Table 2.2. The factors discussed in this chapter that influence a sampling program must be considered along with the information in Table 2.2 when developing a facility sampling program. Many of the analyses listed in the table should actually be tested more frequently based on individual plant conditions. As previously stated, the types of wastewater parameters that must be treated or removed in the plant processes will dictate what process control tests need to be performed. For example, if effluent limitations require total phosphorus removal, total phosphorus should be measured at various stages throughout the plant treatment processes.

For various sizes of activated-sludge plants and the different modes of operation, Table 2.3 lists the recommended process tests (WEF, 1992).

Each treatment facility is slightly different, so there is no one sampling and testing program that can be adapted from one facility to another without some type of modification. The person who establishes the sampling program must take all the factors discussed in this chapter into account to produce a useful and meaningful sampling program.

Table 2.2 Suggested sampling locations, analyses performed, sampling types, and sample frequencies for wastewater treatment unit processes. Each facility should consult its discharge permit for additional parameters and frequency information.

Process	Sampling location	Analysis	Use	Sample frequency	Sample type
Single-stage waste stabilization lagoon	Plant influent	BOD[a]	Plant performance	Weekly	Composite
		TSS[b]	Plant performance	Weekly	Composite
		pH	Process control	Daily	Grab
	Pond	pH	Process control	Daily	Grab
		DO[c]	Process control	Daily	Grab
		Temperature	Process control	Daily	Grab
	Plant effluent	BOD	Plant performance	Weekly	Composite
		TSS	Plant performance	Weekly	Composite
		pH	Plant performance	Daily	Grab
		DO	Plant performance	Daily	Grab
		Fecal coliform	Plant performance	Daily	Grab
		Chlorine residual	Process control	Daily	Grab
Primary treatment	Primary influent	BOD	Plant performance	Weekly	Composite
		TSS	Plant performance	Weekly	Composite
		pH	Plant performance	Daily	Grab
		TKN[d,e]	Plant performance	Weekly	Composite
		Ammonia[e]	Plant performance	Weekly	Composite
		TP[e]	Plant performance	Weekly	Composite
	Primary effluent	BOD	Plant performance	Weekly	Composite
		TSS	Plant performance	Weekly	Composite
		pH	Process control	Weekly	Grab
		DO	Process control	Weekly	Grab
		TKN[e]	Plant performance	Weekly	Composite
		TP[e]	Plant performance	Weekly	Composite
	Primary sludge	TS[f]	Process control	Daily	Composite
		VS[g]	Process control	Weekly	Composite
Trickling filter and rotating biological contactor	Filter influent	BOD	Plant performance	Daily[h]	Composite
		TSS	Plant performance	Daily[h]	Composite
		pH	Process control	Daily	Grab
		DO	Process control	Daily	Grab
	Filter effluent	DO	Process control	Daily	Grab
		pH	Process control	Daily	Grab
		Temperature	Process control	Daily	Grab
		Ammonia[e]	Plant performance	Weekly	Composite
		Nitrate[e]	Plant performance	Weekly	Composite

Table 2.2 Suggested sampling locations, analyses performed, sampling types, and sample frequencies for wastewater treatment unit processes. Each facility should consult its discharge permit for additional parameters and frequency information (continued).

Process	Sampling location	Analysis	Use	Sample frequency	Sample type
Trickling filter and rotating biological contactor (continued)	Final settling tank effluent	BOD	Plant performance	Daily[h]	Composite
		TSS	Plant performance	Daily[h]	Composite
		pH	Process control	Daily	Grab
		DO	Plant performance	Daily	Grab
		Fecal coliform	Plant performance	Daily	Grab
		Chlorine residual	Process control	Daily	Grab
		Ammonia[e]	Plant performance	Weekly	Composite
		Nitrate[e]	Plant performance	Weekly	Composite
		TP[e]	Plant performance	Weekly	Composite
	Final settling tank sludge (secondary sludge)	TS	Process control	Daily	Composite
		VS	Process control	Weekly	Composite
Anaerobic digestion	Digester feed	TS	Plant performance	Daily	Composite
		VS	Plant performance	Daily	Composite
		pH	Process control	Daily	Grab
		Alkalinity	Process control	Twice per week	Grab
	Digester contents	Temperature	Process control	Daily	Grab
		pH	Process control	Daily	Grab
		Volatile acids	Process control	Twice per week	Grab
		Alkalinity	Process control	Twice per week	Grab
		Heavy metals[e]	Process control	Monthly	Grab
	Digested sludge	TS	Plant performance	Daily	Grab
		VS	Plant performance	Daily	Grab
		Volatile acids	Process control	Weekly	Grab
		TKN[e]	Process control	Weekly	Grab
	Supernatant	TSS	Process control	Daily	Composite
		BOD	Process control	Daily	Composite
		Nitrate or ammonia[e]	Plant performance	Weekly	Composite
	Digester gas	Methane or carbon dioxide gas	Process control	Daily	Grab

Table 2.2 Suggested sampling locations, analyses performed, sampling types, and sample frequencies for wastewater treatment unit processes. Each facility should consult its discharge permit for additional parameters and frequency information (continued).

Process	Sampling location	Analysis	Use	Sample frequency	Sample type
Aerobic digestion	Digester feed	TS	Plant performance	Daily	Composite
		VS	Plant performance	Daily	Composite
		pH	Process control	Daily	Grab
		Alkalinity	Process control	Weekly	Grab
		Nitrate or ammonia[e]	Plant performance	Weekly	Grab
	Digester contents	Temperature	Process control	Daily	Grab
		pH	Process control	Daily	Grab
		DO	Process control	Daily	Grab
		TS	Plant performance	Daily	Composite
		VS	Plant performance	Daily	Composite
		Nitrate or ammonia[e]	Process control	Twice per week	Grab
		Alkalinity	Process control	Twice per week	Grab
	Settled digested sludge	TS	Plant performance	Daily	Composite
		VS	Plant performance	Daily	Composite
		pH	Process control	Daily	Grab
		Volatile acids	Process control	Weekly	Grab
		Nitrate or ammonia[e]	Plant performance	Weekly	Grab
	Supernatant	TSS	Process control	Daily	Composite
		BOD	Process control	Daily	Composite
		Nitrate or ammonia[e]	Plant performance	Weekly	Composite
Sludge dewatering and thickening (belt filter press, centrifuge, dissolved air flotation, gravity belt thickener, vacuum filter)	Feed sludge	TS	Plant performance	Daily	Composite
		VS	Plant performance	Daily	Composite
		pH	Process control	Weekly	Grab
	Filtrate, centrate, subnatant	TS	Process control	Daily	Composite
		VS	Process control	Daily	Composite
		pH	Process control	Weekly	Grab
	Dewatered cake	TS	Plant performance	Daily	Composite
		VS	Plant performance	Daily	Composite
		Heavy metals[e]	Plant performance	Monthly	Composite
		Ammonia[e]	Plant performance	Monthly	Composite
Heat drying and incineration	Dewatered cake feed	TS	Plant performance	Daily	Composite
		VS	Plant performance	Daily	Composite
		Temperature	Process control	Weekly	Grab

Table 2.2 **Suggested sampling locations, analyses performed, sampling types, and sample frequencies for wastewater treatment unit processes. Each facility should consult its discharge permit for additional parameters and frequency information (continued).**

Process	Sampling location	Analysis	Use	Sample frequency	Sample type
Heat drying and incineration (continued)	Dried or incinerated solids	TS	Plant performance	Daily	Composite
		VS	Plant performance	Daily	Composite
		Nitrogen[e]	Plant performance	Monthly	Composite
		Phosphorus[e]	Plant performance	Monthly	Composite
		Potassium[e]	Plant performance	Monthly	Composite
	Stack gas	Oxygen	Process control	Weekly	Grab
		Carbon dioxide gas, carbon monoxide gas	Process control	Weekly	Grab
		Particulates	Process control	Weekly	Grab
		Temperature	Process control	Weekly	Grab
		Sulfur dioxide gas	Process control	Monthly	Grab
		Nitrite	Process control	Monthly	Grab
		Hydrocarbons	Process control	Monthly	Grab

[a] BOD = biochemical oxygen demand.

[b] TSS = total suspended solids.

[c] DO = dissolved oxygen.

[d] TKN = total Kjeldahl nitrogen.

[e] TP = total phosphorus; required if discharge requirements dictate (that is, ammonia-nitrogen, phosphorus, nitrogen removal) or if process upsets require investigation.

[f] TS = total solids.

[g] VS = volatile solids.

[h] Frequency could be reduced if the discharge permit allows and/or process upsets are infrequent.

Table 2.3 Recommended process testing for various activated-sludge plants. Each facility should consult its discharge permit for additional parameters and frequency information.

Sample location	Analysis	Frequency	Sample type
Type of plant: activated sludge			
Mode of operation: contact stabilization			
Design flow: 0 to 100 000 gpd[a]			
NPDES permit discharge criteria: BOD,[b] 5–30 mg/L; TSS,[c] 30 mg/L; nutrients, no limit			
Influent	pH	Daily	Grab
	BOD	Weekly	Composite
	TSS	Weekly	Composite
	TKN[d]	Monthly	Grab
	Ammonia	Monthly	Grab
	Alkalinity	Monthly	Grab
Aeration basin	DO[e]	Daily	*In situ*
	Temperature	Daily	*In situ*
Aeration basin effluent	TSS (MLSS[f])	Daily	Grab
	Settleability (5/30/60 minutes)	Daily	Grab
	pH	Weekly	Grab
	Microscopic examination	Weekly	Grab
Return activated sludge flow	TSS	Daily	Grab
	Flow	Daily	Totalizer
Waste activated sludge flow	TSS	Daily	Grab
	Flow	Daily	Totalizer
Secondary clarifier	DOB[g]	Daily	*In situ*
Secondary clarifier effluent	pH	Daily	Grab
	Turbidity	Daily	Grab
	BOD	Weekly	Composite
	TSS	Weekly	Composite
	Ammonia	Monthly	Composite
	Nitrate	Monthly	Composite
	Nitrite	Monthly	Composite

Table 2.3 Recommended process testing for various activated-sludge plants. Each facility should consult its discharge permit for additional parameters and frequency information (continued).

Sample location	Analysis	Frequency	Sample type
Type of Plant: activated sludge			
Mode of operation: contact stabilization			
Design flow: >100 000 gpd			
NPDES permit discharge criteria: BOD, 5–30 mg/L; TSS, 30 mg/L; nutrients, no limit			
Influent	pH	Daily	Grab
	BOD	Weekly	Composite
	TSS	Weekly	Composite
	TKN	Weekly	Grab
	Ammonia	Weekly	Grab
	Alkalinity	Monthly	Grab
Aeration basin	DO	Daily	*In situ*
	Temperature	Daily	*In situ*
Aeration basin effluent	TSS (MLSS)	Daily	Grab
	Settleability (5/30/60 minutes)	Daily	Grab
	pH	Daily	Grab
	Microscopic examination	Weekly	Grab
Return activated sludge flow	TSS	Daily	Grab
	Flow	Daily	Totalizer
Waste activated sludge flow	TSS	Daily	Grab
	Flow	Daily	Totalizer
Secondary clarifier	DOB	Daily	*In situ*
Secondary clarifier effluent	pH	Daily	Grab
	Turbidity	Daily	Grab
	BOD	Weekly	Composite
	TSS	Weekly	Composite
	Ammonia	Weekly	Composite
	Nitrate	Monthly	Composite
	Nitrite	Monthly	Composite
Type of plant: activated sludge			
Mode of operation: conventional or step feed			
Design flow: 1.0 to 10 mgd			
NPDES permit discharge criteria: BOD, 5–30 mg/L; TSS, 30 mg/L; nutrients, no limit			
Influent	pH	Daily	Grab
	BOD	Daily	Composite

Sample location	Analysis	Frequency	Sample type
	TSS	Daily	Composite
	TKN	Weekly	Grab
	Ammonia	Weekly	Grab
	Alkalinity	Weekly	Grab
Aeration basin	DO	Daily	*In situ*
	Temperature	Daily	*In situ*
Aeration basin effluent	TSS (MLSS)	Daily	Grab
	Settleability (5/30/60 minutes)	Daily	Grab
	pH	Daily	Grab
	Microscopic examination	Daily	Grab
Return activated sludge flow	TSS	Daily	Grab
	Flow	Daily	Totalizer
Waste activated sludge flow	TSS	Daily	Grab
	Flow	Daily	Totalizer
Secondary clarifier	DOB	Daily	*In situ*
Secondary clarifier effluent	pH	Daily	Grab
	Turbidity	Daily	Grab
	BOD	Daily	Composite
	TSS	Daily	Composite
	Ammonia	Weekly	Composite
	Nitrate	Weekly	Composite
	Nitrite	Weekly	Composite

Type of plant: activated sludge
Mode of operation: extended aeration
Design flow: 0 to 100 000 gpd
NPDES permit discharge criteria: BOD, 5–30 mg/L; TSS, 30 mg/L; nutrients, no limit

Sample location	Analysis	Frequency	Sample type
Influent	pH	Daily	Grab
	BOD	Weekly	Composite
	TSS	Weekly	Composite
	Alkalinity	Weekly	Grab
	TKN	Monthly	Grab
	Ammonia	Monthly	Grab
Aeration basin	DO	Daily	*In situ*
	Temperature	Daily	*In situ*

Table 2.3 **Recommended process testing for various activated-sludge plants. Each facility should consult its discharge permit for additional parameters and frequency information (continued).**

Sample location	Analysis	Frequency	Sample type
Aeration basin effluent	TSS (MLSS)	Daily	Grab
	Settleability (5/30/60 minutes)	Daily	Grab
	pH	Weekly	Grab
	Microscopic examination	Weekly	Grab
Return activated sludge flow	TSS	Daily	Grab
	Flow	Daily	Totalizer
Waste activated sludge flow	TSS	Daily	Grab
	Flow	Daily	Totalizer
Secondary clarifier	DOB	Daily	*In situ*
Secondary clarifier effluent	Turbidity	Daily	Grab
	pH	Weekly	Grab
	BOD	Weekly	Composite
	TSS	Weekly	Composite
	Ammonia	Weekly	Composite
	Nitrate	Monthly	Composite
	Nitrite	Monthly	Composite

Type of plant: activated sludge
Mode of operation: extended aeration
Design flow: >100 000 gpd
NPDES permit discharge criteria: BOD, 5–30 mg/L; TSS, 30 mg/L; nutrients, no limit

Sample location	Analysis	Frequency	Sample type
Influent	pH	Daily	Grab
	BOD	Weekly	Composite
	TSS	Weekly	Composite
	TKN	Weekly	Grab
	Ammonia	Weekly	Grab
	Alkalinity	Weekly	Grab
Aeration basin	DO	Daily	*In situ*
	Temperature	Daily	*In situ*
Aeration basin effluent	TSS (MLSS)	Daily	Grab
	Settleability (5/30/60 minutes)	Daily	Grab
	pH	Daily	Grab

Table 2.3 Recommended process testing for various activated-sludge plants. Each facility should consult its discharge permit for additional parameters and frequency information (continued).

Sample location	Analysis	Frequency	Sample type
	Microscopic examination	Weekly	Grab
Return activated sludge flow	TSS	Daily	Grab
	Flow	Daily	Totalizer
Waste activated sludge flow	TSS	Daily	Grab
	Flow	Daily	Totalizer
Secondary clarifier	DOB	Daily	*In situ*
Secondary clarifer effluent	pH	Daily	Grab
	Turbidity	Daily	Grab
	BOD	Weekly	Composite
	TSS	Weekly	Composite
	Ammonia	Weekly	Composite
	Nitrate	Weekly	Composite
	Nitrite	Weekly	Composite

[a] $gpd \times (3.785 \times 10^{-3}) = m^3/d$.

[b] BOD = biochemical oxygen demand.

[c] TSS = total suspended solids.

[d] TKN = total Kjeldahl nitrogen.

[e] DO = dissolved oxygen.

[f] MLSS = mixed liquor suspended solids.

[g] DOB = depth of blanket.

REFERENCES

American Public Health Association (1995) *Standard Methods for the Examination of Water and Wastewater.* 19th Ed., Washington, D.C.

Kopp, J.F., *et al.* (1983) *Methods for Chemical Analysis of Water and Wastes.* EPA-600/4-79-020, U.S. EPA, Washington, D.C.

U.S. Environmental Protection Agency (1989) *POTW Sludge Sampling and Analysis Guidance Document.* Washington, D.C.

Water Environment Federation (1992) Activated Sludge: Improve Testing to Improve Performance. *Operations Forum,* **9,** 8.

SUGGESTED READINGS

Sacramento State College (1993) *Operation of Wastewater Treatment Plants: A Field Study Training Program.* Sacramento, Calif.

U.S. Environmental Protection Agency (1978) *NPDES Compliance Sampling Inspection Manual.* MCD-51, Washington, D.C.

U.S. Environmental Protection Agency (1992) *Environmental Regulations and Technology: Control of Pathogens and Vectors in Sewage Sludge.* EPA-1625/R-92-013, Washington, D.C.

Water Environment Federation (1990) *Operation of Wastewater Treatment Plants.* Vol. 2, Manual of Practice No. 11, Alexandria, Va.

Chapter 3
Sample Collection Equipment

INTRODUCTION

The most accurate data on a waste stream would be obtained by collecting and analyzing the entire flow from a discharge point. Because this is not practical, sampling devices are used to collect and store aliquots that accurately reflect the composition of the source. These devices must collect samples that are of sufficient volume to represent the source but small enough to be handled in the laboratory. Samples should also be collected in a manner that reflects the concentrations of pertinent constituents in the total discharge. After the samples are collected, they must be preserved and handled properly so that no significant change in the samples can occur before analysis.

Sampling equipment is but one component of an overall sampling plan. The sampling plan is used by professionals to detail all requirements for gathering data in a particular sampling location. This plan outlines the correct method for determining the equipment and procedures needed for sample collection.

NEED FOR EQUIPMENT. Although a manually collected sample can accurately reflect the source in question, there are many situations in which manual sampling is not practical, such as when samples must be collected continuously, from hazardous locations, or when personnel are not available to perform the sampling. These are situations in which automatic sampling equipment is advantageous.

In addition, the accuracy of manual samples depends on the skill of the person collecting them. Individual aliquots collected by different people from the same source may reflect widely varying constituents. Automatic samplers collect samples with the same method, from the same point, at programmed intervals. This uniformity ensures that samples delivered for analysis are consistent and accurately reflect the conditions in the source.

The correctly collected manual sample, nevertheless, is still the basis of comparison for samples collected automatically. The proper procedures and equipment used to collect representative samples manually are discussed below.

PERSONNEL SKILLS AND TRAINING. Careful precision in the laboratory cannot compensate for improperly collected samples: the quality of the data generated largely depends on the quality of the samples collected. In fact, errors caused by inadequate field techniques, poor equipment selection, and improper sample preservation will have a greater effect on the testing outcome than errors from variations in laboratory analyses.

For this reason, it is necessary to assess the personnel skills and training resources available for the monitoring program. The need for quality in the sample collection process, ensured through proper training, is important to obtaining accurate information on the data source. Proper training and technique also result in better decisions about pollution control required for a particular discharge. Representative samples, properly collected and preserved, are worth the effort in the quality of data they produce.

Training personnel responsible for sample collection is critical to the success of any sampling program. Individuals need to understand the sampling plan and the goals of the program. Employee skills must be assessed and individuals made aware of the importance of their efforts. A detailed training program should be developed based on the overall sampling plan.

MAINTENANCE DOWNTIME ALLOWANCES. Once the scope of the monitoring program and equipment have been identified, it is important to estimate the ramifications of equipment downtime. Repair times vary depending on the specific equipment. Plant process and regulatory monitoring requirements, facility purchasing procedures, and the need for continuous data collection warrant a certain level of backup. Finally, a standard operating procedure for interim manual sample collection should be in place if replacement equipment is not readily available.

After these issues have been addressed, it is possible to select equipment by applying the general guidelines provided later in this chapter. If equipment is to be purchased from more than one vendor, care should be taken to ensure that the equipment is compatible.

*M*ANUAL SAMPLE COLLECTION

Sampling of toxic pollutants necessitates specific contact material protocols to prevent degradation of the samples. Sampling equipment and containers should be made of inert materials and should be thoroughly cleaned. Using approved dipping devices from established industry vendors will add to the quality of the samples.

TYPES OF SAMPLERS. Table II of 40 CFR section 136.3 defines specific materials, preservation techniques, and holding times for pollutants. For sampling of priority pollutants, the recommended primary sample container is a 1-L, wide-mouth glass jar with a polytetrafluoroethylene lid liner. For small samples, a 1-pint (0.5-L), wide-mouth glass jar is recommended. If glass is not required, nonbreakable containers should be used. However, the

---

overriding factor in equipment selection is the type of pollutants being monitored. All sampling equipment, including containers, should be thoroughly cleaned before and after each field use. Cleaning of equipment in the field is not recommended.

There are three general classes of manual discrete sampling equipment that are versatile, inexpensive, and simple to use.

Weighted Bottle Sampler. A weighted bottle sampler consists of a bottle (usually glass), a weighted sinker, a bottle stopper, and a line used to lower and raise the container. Several variations are available from major scientific supply houses. Additionally, numerous local or shop-fabricated types can be seen at individual wastewater treatment plants (Shelley, 1977).

These weighted samplers can be used in basins and many process units. They are also used in streams and ponds that are deep enough and have a low enough stream velocity to allow accurate placement of the bottle.

Pond Sampler. The pond sampler consists of an adjustable clamp attached to the end of a two- or three-piece telescoping aluminum tube, which serves as a handle. The clamp secures a sample beaker or bottle. This type of sampler can be made easily and inexpensively. The tubes can be purchased from most hardware or swimming pool supply stores, and the adjustable clamp and sampling containers can be purchased from most laboratory supply houses. Pond samplers can collect liquid samples from disposal ponds, pits, lagoons, and similar reservoirs.

Coliwasa Sampler. The Coliwasa sampler (*Composite liquid waste sampler*) is primarily applicable to the sampling of liquid wastes in containers. Its design permits representative sampling of a wide range of multiphase wastes. The first Coliwasa was designed and built by the California Department of Health. Following its use there, the University of Southern California and the U.S. Environmental Protection Agency (U.S. EPA) developed the current configuration (Figure 3.1).

The sampler's main parts include a positive, quick-engaging, closing mechanism and a sampling tube. The sampling tube is made of borosilicate glass with a polytetrafluoroethylene stopper rod.

Other Types of Samplers. Other samplers are available for specific applications. Items such as the Van Dorn and Kemmerer samplers and biochemical oxygen demand (BOD) bottle samplers (APHA, 1995) can provide quality manual grab samples.

Sampling for volatile organic compounds (VOCs) is predominantly a manual task. Sampling for VOCs requires special protocols to collect and store the samples for proper analysis. Federal regulation 40 CFR Part 136, App A, Method 602, states "Fill the sample bottle in such a manner that no air bub-

---

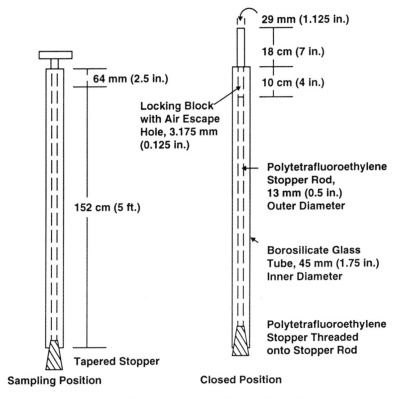

Figure 3.1 Composite liquid waste sampler (Coliwasa).

bles pass through the sample as the bottle is being filled. Seal the bottle so that no air bubbles are entrapped in it. Maintain the hermetic seal on the sample bottle until time of analysis." This regulation also states that the samples must be kept on ice or refrigerated until analysis.

Typically, because of these strict protocols, this type of sampling is performed only periodically, and the samples are collected manually. However, some new advances in this area of sampling are discussed later.

*A*UTOMATIC SAMPLE COLLECTION EQUIPMENT

There are more than 100 different automatic sampling devices available commercially for collecting wastewater. Additionally, there are many custom-built, single-purpose automatic devices for special sampling applications. These devices might be used to sample sources consisting of sludge, high suspended solids, extreme depths, or difficult-to-reach collection points.

Nevertheless, no single sampler will provide acceptable samples for every possible field condition.

EVALUATING AUTOMATIC SAMPLER SYSTEMS. Some desirable characteristics of automatic compositors are discussed by Shelley and Kirkpatrick (1973). This report is an effective framework for developing a sampling plan and determining automatic equipment requirements. Not all of the desirable features listed are applicable to every application. The features that best meet the objectives of a specific sampling plan will guide the selection of automatic equipment.

Intake Device. The intake device is the liquid interface for the automatic sampler (Figure 3.2). Because it is typically submerged in the flow stream, the appropriate intake device should

- Cause minimal obstruction in the sewer or channel. This will decrease the possibility of damage to or fouling of the intake.
- Be capable of drawing a sample from the entire stream, including the surface, middle, and bottom layers (Rabosky and Koraldo, 1973). In cases where the flow stream is highly stratified or relatively deep, multiple sampling points may be required.
- Be adaptable to a variety of sampling applications. To meet this requirement, nozzles, weighted vertical lines, and taps on pressure lines may be needed. In general, screen intakes are undesirable because they may not provide representative samples.
- Have an internal diameter small enough to ensure the desired line velocities but large enough to prevent clogging and collect suspended solids of interest. The U.S. EPA-recommended internal diameter is at least 6.4 mm (0.25 in.).

Sample Transport. The sample transport is the tubing path from the source to the storage container. The major concern is the material used for the path. The sample collection protocols will determine the material acceptable for specific pollutants. In general, the transport line should exhibit the following characteristics:

- Disallow metals contact during sample collection.
- Prevent obstructions in the sample line, minimizing cleaning problems.
- Consist of flexible tubing made of transparent, inert material. If measurements are conducted for trace quantities of a pollutant, a method for testing the tubing for contamination is needed (Junk *et al.*, 1974).

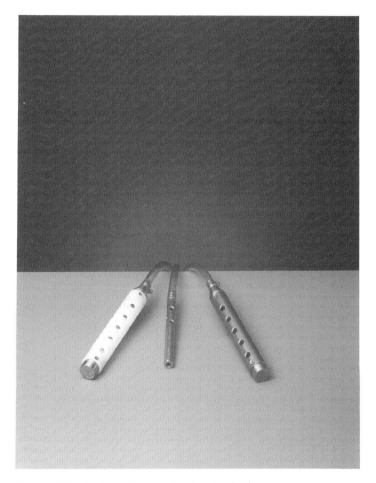

Figure 3.2 Automatic sampler intake device.

Sample Collection (Pumping System). A common type of sampling pump used in the U.S. is the suction lift pump. These pumps are available as high-speed peristaltic type pumps and vacuum lift pumps. Some of the desirable features include

- Pump controls that allow purging and rinsing of the transport path. A clean water purge is desirable, but it is not feasible in most applications. A complete air purge is sufficient for most sampling.
- Ability to pump dry without damage.
- Self-priming pump.
- No metals contact with the sample.

- A variable intake flow velocity between 0.60 and 3.0 m/s (2.0 and 10.0 ft/sec). In isokinetic sampling, the intake velocity of the pump matches the velocity of the flow stream being sampled, allowing for more representative solids collection. U.S. EPA recommends a sample line velocity of 0.60 to 0.75 m/s (2 to 3 ft/sec) for most sampling applications.
- Pump capacity to lift samples for a head (vertical distance) of at least 6 m (20 ft). In practice, however, the vertical lift of the samples should be kept to minimum.
- Pump capacity to maintain line velocity of 0.6 to 3.0 m/s (2 to 10 ft/sec) (Rabosky and Koraldo, 1973).
- Sample volume independent of head height.
- Pump and motor assembly enclosed in a water-tight container. This protects the motor from water in the event of a pump leak.

Power and Controls. The following features should be available in sampler controllers to allow for adaptability to a variety of conditions:

- Capability for alternating current (ac) and direct current (dc) operation.
- Battery life of a minimum of 5 days of hourly sampling without recharging.
- Battery weight of less than 9 kg (20 lb).
- Sealed battery to prevent leaks.
- Solid-state logic and printed circuit boards.
- Timing and control systems contained in a water-proof compartment for humidity and submergence protection.
- Solid-state controlled electronics.
- Ability to accommodate both flow-proportional sampling (directly linked to a flow meter) and time-based sampling. The timing interval should allow for a range of 10 minutes to 4 hours.
- Adaptability to different sample distribution schemes. The controller should be capable of placing multiple samples in the same container, to produce "miniature" composite samples, or placing samples in multiple bottles, for situations involving different preservatives or requiring duplicate samples.

SAMPLE STORAGE. Proper storage is integral to maintaining representative collections of samples. Different sampling sites require different storage protocols. The automatic sampler should

- Collect discrete samples in multiple containers. Additionally, the use of a single composite container should be available.
- Contain discrete sample containers at least 400 mL in size. Single composite containers should be at least 9.4 L (2.5 gal).

- Hold at least 24 discrete sample containers.
- Consist of wide-mouth containers made of chemically resistant plastic (such as polyethylene or polypropylene) or borosilicate glass (for soluble oil samples or priority pollutants).
- Have the capability for cooling samples via refrigeration or packed ice.
- Have an insulated sample storage compartment. This compartment protects samples in cold weather and increases the efficiency of the ice used for cooling in warm weather.

OTHER DESIRABLE FEATURES. If the automatic sampler is being used to collect samples from several locations, various factors must be taken into consideration. The following items may be needed for some sampling operations. Keep in mind that certain trade-offs may be required for the unit to operate in all conditions:

- Watertight casing to protect the internal components from submergence and high-humidity conditions;
- Vandalproof casing and sampler that can be locked;
- A secure harness or mounting device, if sampler is placed in a sewer;
- Explosionproof construction (Developing Water, 1974);
- Sizing that allows sampler to fit into a standard manhole without disassembly;
- Construction that is compact, portable, and light enough to allow one-person manhole installation (in accordance with Occupational Safety and Health Administration requirements for manhole entry); and
- Construction (including casing) of corrosion-resistant materials such as plastics, fiber glass, and stainless steel.

INSTALLATION AND USE OF AUTOMATIC WASTEWATER SAMPLERS. Well-designed sampling equipment will elicit good results only if it is properly maintained and applied (Harris and Keffer, 1974). When the sampler is installed in a manhole, it should be secured to a rung in the manhole or, if space permits, placed at the bottom of the manhole. If the sampler is outside the manhole, additional security may be required. For example, the equipment could be locked and secured to a stake. The sampler should not be placed in an area where it could be submerged. Installation below ground reduces sample suction head, eliminates vandalism, and provides a more uniform operating temperature.

The intake tubing should either be placed vertically or sloped to ensure gravity drainage of the tubing between samples. Loops or dips in the line can become areas of cross-contamination or line blockage in the winter and should be avoided.

Sample bottles, tubing, and any portion of the sampler that contacts the sample should be cleaned after each use. Whatever the cleaning methods, all sampler parts that come in contact with the sample should be given a final rinse with both tap water and distilled water for best results.

The sampler intake should be inspected after each use and cleaned if necessary. To ensure a representative suspended solids sample, the intake(s) should be carefully placed. If the intake is allowed to rest at the bottom of a flow stream, the intake purge may dislodge solids that are not part of the discharge. The intake should also be facing upstream. This position is best for taking representative samples, although at times it can cause fouling and plugging of the intake. It is important that the intake be in a turbulent area where the flow is well mixed to help reduce plugging of the intake. The intake should be secured at all times, with no drag on the inlet tubing. For a single intake, the sample should be collected at a point of maximum velocity, slightly below the center of the water depth.

Electrical and mechanical parts should be maintained according to the manufacturer's instructions. Typically this entails only minimal preventive maintenance. The desiccant should be replaced as needed. If a wet-cell, lead-acid battery is used, any spilled acid should be cleaned and neutralized. In peristaltic pump samplers, the tubing used inside the pump will need to be maintained in accordance with the manufacturer's instructions.

WINTER OPERATION. In freezing outdoor temperatures, special precautions should be taken to ensure reliable sample collection and prevent collected samples from freezing. The sampler should be placed below the freezing level or in an insulated box.

When ac line current is available, a light bulb or heating tape can be used to warm the sampler, if required. The following arrangement has been found to be satisfactory between –18 and –12°C (0 and 10°F) (Developing Water, 1974). A short 1.2- or 1.8-m (4- or 6-ft) thermostatically protected 3°C (38°F) heat tape is wrapped around the sample bottle and the intake lines on the ac samplers. Over the heat tape on the intake, a large plastic bag (5 mL thick) is loosely wrapped. A large plastic bag also can also be placed over the sampler.

Other solutions to the need for suction line insulation can range from heated tubing to simply wrapping insulation around the line. The knowledge of the sampling site will dictate the precautionary measures needed for continuous operation.

Newer refrigerated samplers include built-in heaters that eliminate the need for protection of the components, except for the suction line. These units typically have devices to heat the sampler and prevent freezing of samples under most conditions. The suction line, if not properly routed, could become blocked by ice in the line.

The actual suction line length should be kept to a minimum. Additionally, the suction line should be either placed vertically or sloped to the source. This configuration will provide a gravity drain back to the source to remove any liquid left after purging of the suction line.

Buildups in the sample line may still occur during sampling of some flow streams. The entire sampling system should be placed on a preventive maintenance schedule consistent with the site conditions. Some sites may require frequent changing of pump and/or suction line tubing or other simple maintenance considerations.

TYPES OF AUTOMATIC SAMPLERS AND THEIR USE

PORTABLE SAMPLING EQUIPMENT. Equipment suppliers can provide portable sampling equipment that is easily transported to various field sites. This equipment can be ordered in versions that conveniently fit into manhole structures and other small sampling areas. This equipment is generally convertible to single-bottle composite or multiple-bottle configurations. Bottles are available in various sizes and materials to accommodate various sample collection protocols.

Programmable samplers can collect samples at specified time intervals, or they can be linked to flow signals for flow-proportional sampling. Some models can be activated by liquid-level or other parameter sensors such as pH or dissolved oxygen sensors. The new samplers include data-logging capabilities for storing information regarding each sample collection and other external parameters. Typically, portable samplers are required to operate in less than ideal conditions and must be uniquely adapted to a specific situation. These units can be connected to accessory components to provide solar power capabilities, special installation harnesses, computer and telephone interface connections, or other specialized sampling needs.

REFRIGERATED SAMPLING EQUIPMENT. For permanent sampling needs, temperature-controlled equipment is available to ensure sample integrity. This is especially pertinent to applications in which temperature preservation is required, such as testing for BOD_5. These units typically are ac-powered and have internal temperature controls, allowing operation in conditions where temperatures can fluctuate from -20 to $120°C$.

Refrigerated samplers typically accommodate larger composite containers than portable samplers. Most are adaptable to discrete bottle sampling in a variety of bottle configurations.

SPECIAL SAMPLING EQUIPMENT. Many special-purpose devices have been adapted to the particular requirements of different sampling sites. These devices are designed to solve some type of operational or application problem at the site.

Flow-Through Samplers. Flow-through samplers are most often refrigerated units used in permanent applications. This device typically will use some type of pump to route part of the flow stream being sampled through small-diameter piping (usually 50-mL [2-in.] line) to the sampler. The sampler has adaptors for the small piping. The pumped flow is directed inside the sampler to a chamber, where one of two methods is used to collect the sample: a "dipper" mechanism or a suction pump. If the samples are collected using a dipper mechanism, a weir plate constructed inside the chamber creates a pool of water for the mechanical device to dip in and collect the sample. If a suction pump is used, no weir plate is required, and the intake to the pump is located in the flowing water. The excess water from the stream is returned to the source by gravity or other pumped means.

The advantage of a flow-through sampler is that the sample retrieval point does not need to be near the sample collection point. Samples can be pumped from specific locations in a treatment plant to a point near the laboratory for convenient retrieval.

Automatic Samplers for Volatile Organic Compounds. Until recently, samples analyzed for VOCs were collected manually. As discussed previously, this can lead to errors in the sample accuracy because of variations in the collection technique or human errors.

In the past 2 years, new automatic samplers have been introduced to the market that allow a user to automatically collect samples for VOCs. Unique collection methods are needed to comply with federal requirements for sample collection.

Some of these units operate in closed-pipe flow situations only. Samples are trapped in a section of the pipe, separate from the flow stream. This type of system would have applications in the process monitoring area of sampling. Other units can collect samples from open channels and closed conduits. One of these units collects samples in standard 40-mL vials with special valves.

Pressurized Line Samplers. Often, samples are required in closed-channel applications. The type of sample and frequency of sampling required will determine the equipment necessary.

In some locations, a simple valve tap can be placed on the line to allow a person to open a faucet and manually collect a sample. If the pressurized line does not exceed 100 kPa (15 psi), typical peristaltic pump samplers can be used to collect samples automatically.

If pressurized lines exceed 100 kPa (15 psi), specialized sampling devices may be required. An example of a special device might be a device used for opening a solenoid valve to activate a sample collection. A dedicated device that rotates into the flowing line and fills rapidly might also be used. After filling, the device rotates out of the flow stream and the sample is deposited in a container.

Other Equipment. On-line monitoring equipment is rapidly being introduced to the field by several manufacturers. This equipment includes on-line total organic carbon (TOC), BOD, and total oxygen demand monitoring equipment.

Total organic carbon monitoring has been shown to be a good indicator of overall water quality. The TOC parameter is commonly tested in laboratories and has more recently been applied to industrial and municipal treatment processes in the field. In addition to providing continuous information about water quality, the on-line TOC analyzer can be used for spill or leak detection at certain facilities. Other on-line analyzers will be adapted to field use as improved technology allows the equipment to adequately withstand the rigors of field applications.

FLOW MEASUREMENT WITH WASTEWATER SAMPLING

OPEN-CHANNEL FLOW MEASUREMENT. The measurement of flow in conjunction with water quality sample collection is essential. As Kirkpatrick and Shelley (1975) state, "Measurements of quantity of flow, usually in conjunction with sampling for flow quality, are essential to nearly all aspects of water pollution control. Research, planning, design, operation and maintenance, and enforcement of pertinent laws—all are activities which rely on flow measurement for their effective conduct."

Flow data should be collected with the same care and precision required for representative sample collection. Uniform and reliable measurement data are needed to identify the resource levels and quality in bodies of water for determining the results of conservation and quality control efforts and enforcing water conservation and regulatory requirements.

Flow data may be collected on an instantaneous or continuous basis. Instantaneous flow rate data can be obtained using several methods, including the timed gravimetric method, dye dilution, or instantaneous velocity measurement.

An automatic flow-measurement system is required for collecting continuous data. Automatic flow-measurement systems are also preferred for collecting samples according to the flow rate. A typical system consists of the primary measurement device (usually a hydraulic structure with a known level-to-flow relationship), a level sensor, a control and conversion device, a recorder, and a totalizer. Many automatic systems are compatible with portable personal computers for downloading stored data from remote field sites.

Additional options can include parameter measurements, telemetry options, and other devices needed for specific sites.

Primary Measuring Devices. As discussed in the previous section, the most commonly used method of measuring flow rate in an open channel is the use of hydraulic structures. In this method, flow in an open channel is measured by inserting a hydraulic structure into the channel, which changes the level of flow in or near the structure. Based on the shape and dimensions of the hydraulic structure, the rate of flow through or over the restriction can be related to the liquid level. In this way, the flow rate through the open channel can usually be derived from a single measurement of the liquid level. The hydraulic structures used in measuring flow in open channels are known as primary measuring devices and may be divided into two broad categories: weirs and flumes.

WEIRS. A weir (Figure 3.3) is essentially a dam, built across an open channel, over which liquid flows, usually through an opening or notch. Weirs are normally classified according to the shape of the notch, the most common types being the rectangular weir, the trapezoidal (or Cipolletti) weir, and the triangular (or V-notch) weir. For each type of weir, an associated equation can be used to determine the flow rate over the weir, allowing a measurement of the liquid level in the upstream pool to accurately reflect the flow rate in the channel.

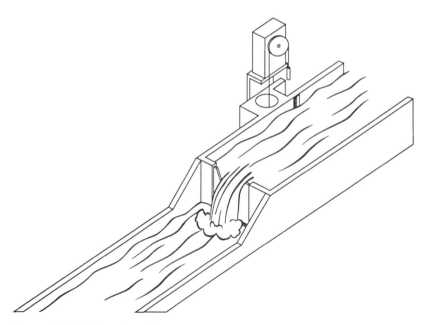

Figure 3.3   Weir schematic.

Although there are many different types of weirs, they all share several common characteristics (Figure 3.4). The *crest* of the weir is the actual point at which water flows over the structure. (The triangular weir ends in a point on the weir plate and, therefore, technically has no crest length. However, the point at which water flows over the structure is referred to as the crest.)

The *nappe* of the weir is the flow stream leaving the hydraulic structure. For proper operation, this stream of water should spring past and not cling to the weir plate. When the downstream water level rises to the point at which air fails to flow freely beneath the nappe, the low pressure in this area may cause an inaccurate discharge rate reading.

The discharge rate of a weir is determined by measuring the vertical distance from the crest of the weir to the liquid surface in the pool upstream from the crest. This liquid depth is called the *head*. As shown in Figure 3.4, the liquid surface begins to drop slightly upstream from the weir. This drop begins at a distance of at least twice the head on the crest and is called the surface contraction, or *drawdown*, of the weir. To avoid measuring the effects of drawdown, the head measuring point of the weir should be located upstream of the weir crest a distance of at least three, and preferably four, times the maximum head expected over the weir, as shown in Figure 3.4. If the maximum head height is not known, the measuring point should be three to four times the head height the weir is capable of measuring. Once the head is known, the flow rate or discharge can be determined using the known head-to-flow rate relationship of the weir.

During sampling of a flow stream over a weir, the sample intake should be on the downstream side of the weir plate. These samples should be collected

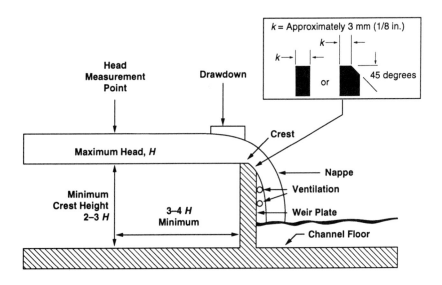

Figure 3.4   Characteristics of a weir.

some distance beyond the turbulent area of discharge from the weir. This area of turbulence provides a thoroughly mixed sample. If samples are collected in an area before the weir plate, stratification of pollutants can be caused by the weir's presence. This can lead to samples that do not adequately represent the flow stream.

Many weir configurations have been used in the past to account for specific flow situations. However, most weirs in use today are either triangular, rectangular, or trapezoidal.

Triangular (V-Notch) Weirs. The V-notch weir may be the most popular weir in use today (Figure 3.5). The V-notch weir consists of an angular notch cut into a bulkhead in the flow stream. The angle of the notch most often used is 90 deg. However, V-notch weirs with angles of 22.5, 30, 45, 60, and 120 deg are also common. The V-notch weir is an accurate flow-measuring device particularly suited to low flows. Because the V-notch weir has no crest length, the head required for a small flow through it is greater than that required for other types of weirs. This is an advantage when measuring small discharges because the nappe can spring free of the crest, whereas it would cling to the crests of other types of weirs, reducing the accuracy of the measurement.

Rectangular (Contracted and Suppressed) Weirs. The rectangular sharp-crested weir (Figure 3.6) can be used in one of two configurations. The first configuration consists of a rectangular notch cut into a bulkhead in the flow channel, producing a box-like opening. This configuration is called a contracted rectangular weir because of the curved flow path or contraction that results, with the nappe forming a jet narrower than the weir opening. The horizontal distances from the end of the weir crest to the side walls of the

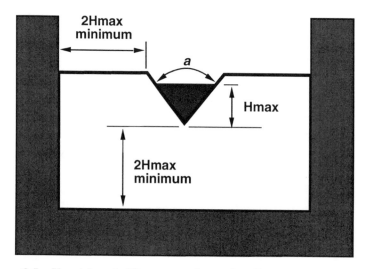

Figure 3.5   V-notch weir (H_{max} = maximum head).

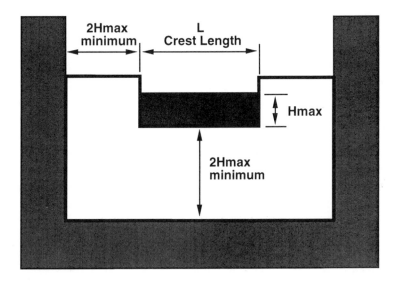

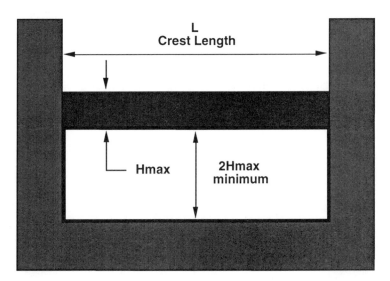

Figure 3.6 Rectangular sharp-crested weir (top, contracted [with end contractions]; bottom, suppressed [without end contractions]) (H_{max} = maximum head).

channel are called the *end contractions*. These end contractions reduce the width of the channel and accelerate the channel flow as it passes over the weir, providing the needed ventilation.

In the second rectangular weir configuration, the end contractions are completely suppressed by extending the weir across the entire width of the channel. Thus, the sides of the channel act as the sides of the weir, and there are no lateral contractions. This type of weir is called a *suppressed* rectangular weir, and flow through it is said to have no end contractions.

Trapezoidal (Cipolletti) Weirs. The trapezoidal sharp-crested weir (Figure 3.7) is similar to a rectangular weir with end contractions except that the sides incline outwardly, producing a trapezoidal opening. When the end-inclinations of a trapezoidal weir are in a ratio of 4:1 with respect to vertical and horizontal slopes of the upstream face respectively, the weir is known as a *Cipolletti* weir. Although the Cipolletti weir is a contracted weir, its discharge occurs essentially as though its end contractions were suppressed. Thus, unlike a rectangular contracted weir, no correction is necessary for the crest width, and the discharge equation is simpler. This type of weir generally is less accurate than a V-notch and does not accommodate the capacity of a rectangular weir installed in a similar-size channel.

OPEN-FLOW NOZZLES. The open-flow nozzle is a combination of a sharp-crested weir and a flume. Flow nozzles are designed to be attached to the end of a conduit flowing partially full and discharging to a free fall. As with weirs, the flow nozzle is used to establish a relationship between the depth of the liquid within the nozzle and the rate of flow.

Two flow nozzle designs are shown in Figure 3.8. In one design, the Kennison nozzle, the cross section is shaped so that the relationship is linear. In the second design, the parabolic nozzle, the parabolic shape causes each unit increase in the flow to produce a smaller incremental increase in head.

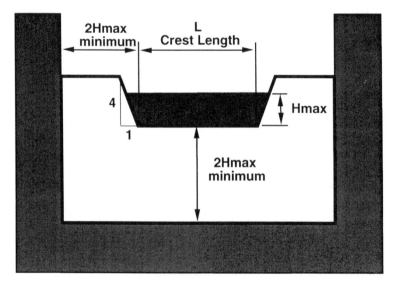

Figure 3.7   Trapezoidal sharp-crested weir (H_{max} = maximum head).

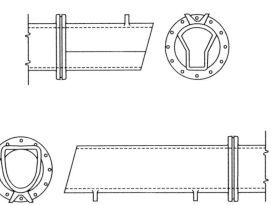

Figure 3.8   Two flow nozzle designs (top, Kennison nozzle [Q proportional to H]; bottom, parabolic nozzle [Q proportional to H^2]).

Open-flow nozzles are factory calibrated and offer reasonable accuracy even under rather severe field conditions. Kennison nozzles are available in standard sizes 150 to 910 mm (6 to 36 in.), with maximum capacities as large as 880 L/s (31 cfs). Parabolic nozzles are available in standard sizes 150 to 610 mm (6 to 24 in.), with maximum capacities as large as 450 L/s (16 cfs).

Unlike conventional weirs, flow nozzles can handle suspended solids rather effectively: a self-scouring action allows relatively large solids to pass without clogging. Use of flow nozzles for heavy sludge is not recommended because deposits will alter the contour of the nozzle and, hence, flow characteristics. The flow nozzle does not feature the low head-loss characteristics of a flume. The loss of head through the device will be at least one pipe diameter because of the restriction in the pipe cross-sectional area caused by the nozzle.

FLUMES. A *flume* (Figure 3.9) is a specially shaped open-channel flow section with an area or slope different from that of the channel. This different area or slope causes increased velocity and a change in the level of the liquid flowing through the flume. A flume normally consists of a converging section, a throat section, and a diverging section (Figure 3.10). Flow rate through the flume is a function of the liquid level at a point or points in the flume. The flow rate through the flume is determined by measuring the head on the flume at a single point, typically some distance downstream from the inlet. The head–flow rate relationship of a flume can be defined by either test data (calibration curves) or by an empirically derived formula.

When a flume is used as the primary measuring device in an open channel, the best sampling site is the diverging section of the flume. This section, which follows the throat section, is the area in which the flow will be most turbulent and best mixed. Several flume manufacturers will offer sampling port installation on permanent flumes as an option.

Sample Collection Equipment   **47**

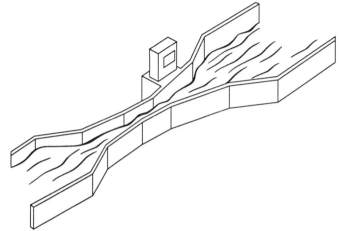

Figure 3.9   Typical flume.

The most commonly used flumes are Parshall and Palmer-Bowlus flumes, although there are many other types available.

Parshall Flume. The Parshall flume (Figure 3.11) is perhaps the most commonly used flume in the wastewater treatment industry. The Parshall flume operates on the principle that open-channel flow, when passing through an exactly prescribed convergence and constriction in the channel, will produce a hydraulic head at a specific point upstream of the constriction. The converging upstream portion of the flume accelerates the entering flow, helping to eliminate deposits of sediment that would otherwise reduce measurement and sampling accuracy.

The flume must remain level and can be affected by upstream approach conditions. A level converging section of the flume is followed by a downward sloping floor at the throat. The flow, which is proportional to the head, is raised to an exponent and multiplied by a throat coefficient. Because it is self-scouring, the Parshall flume is a practical device for determining open-channel flow. The accuracy of the Parshall flume is approximately 5% of the actual

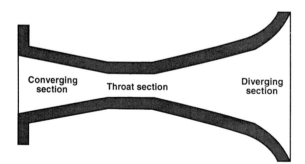

Figure 3.10   Sections of a flume.

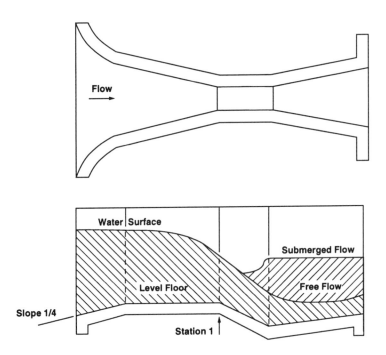

Figure 3.11   Parshall flume (top, plan view; bottom, elevation view).

rate of flow depending on fabrication, installation, and upstream abnormalities.

Some practitioners have used a simplified version of the Parshall flume, sometimes referred to as the *Montana* flume. The crest of a Montana flume should be set above the channel bottom to ensure that the flow profile over the crest section is not modified by backwater from the downstream channel. As long as the 70% submergence limit is not exceeded, the standard discharge equations for Parshall flumes may be applied to similarly sized Montana flumes.

Palmer-Bowlus Flume. One of the primary benefits of the Palmer-Bowlus flume (Figure 3.12) is that it can be placed into an existing channel with either a round or rectangular cross section. The flume is essentially a restriction in the channel designed to produce a higher velocity critical flow in the throat section.

Several factors are important for accurate flow measurement when using the Palmer-Bowlus flume. The upstream channel must be tranquil and smooth with no aeration or surface waves. The water should pass through the throat with little turbulence, allowing the surface profile to drop smoothly

---

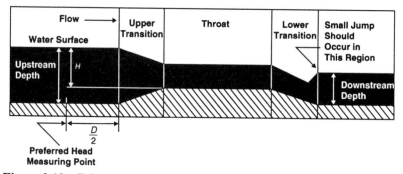

Figure 3.12　Palmer-Bowlus flume, elevation view (*D* = conduit diameter).

throughout the length. A shooting flow downstream of the flume indicates that wastewater is discharging freely. If there is a hydraulic jump, it should begin outside of the flume section.

Figure 3.13 shows three different types of Palmer-Bowlus flumes. Unlike the Parshall flume, the dimensional configuration of a Palmer-Bowlus flume is not rigidly established for each flume size.

A Palmer-Bowlus flume size is identified by the size of the pipe or conduit into which it fits rather than by the throat width, as is the case with Parshall flumes. Thus, a 0.2-m (8-in.) Palmer-Bowlus flume is designed for placement in a 0.2-m (8-in.) diameter pipe. Standard Palmer-Bowlus flumes are available to fit pipe sizes ranging from 0.1 m (4 in.) to 1 m (42 in.) and are typically purchased prefabricated from the flume manufacturer. They are normally made of fiber glass, a reinforced plastic, or stainless steel.

When choosing the size of Palmer-Bowlus flume to be installed at a particular location, both the diameter of the conduit into which the flume is to be installed and the range of expected flow rates should be considered. However, future expansion and the capacity to measure higher expected flows must be considered.

HS, H, HL Flumes. The HS, H, and HL flumes were developed by the U.S. Department of Agriculture Soil Conservation Service in the 1930s to measure small watersheds. They have been successfully used in the wastewater industry to measure surface runoff and wastewater flows (Figures 3.14 and 3.15). Their wide measurement spans make them particularly suitable for measuring drainage water and for portable applications during which a wide range of flow rates may be encountered. Because these flumes require a free discharge, some amount of head loss may result.

Trapezoidal Flumes. The trapezoidal flume was developed primarily to measure flow in irrigation channels and has been used for many years by the Agricultural Research Service, U.S. Department of Agriculture (Figure 3.16). For agricultural applications, the trapezoidal flume is superior to the Parshall

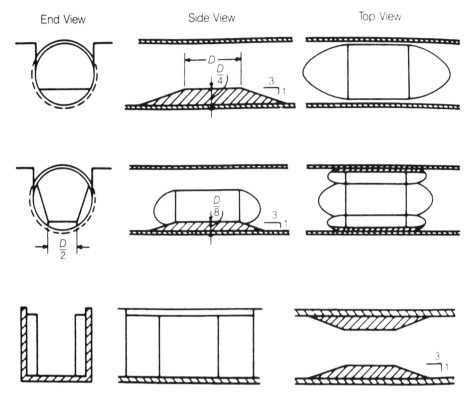

Figure 3.13   Three shapes of Palmer-Bowlus flumes (D = conduit diameter).

flume for a number of reasons, particularly when measuring smaller flows. The trapezoidal shape conforms to the normal shape of ditches, especially those that are lined. This minimizes the amount of transition section needed compared to that required when changing from a trapezoidal shape to a rectangular shape and back to trapezoidal. The trapezoidal shape is also desirable because the side walls expand as depth increases. This means that a trapezoidal flume can convey a larger range of flow rates because, as a result of the trapezoidal shape, an incremental increase in flow produces a relatively small increase in depth.

Trapezoidal flumes should generally be installed level. Occasionally, flumes are installed with a slight slope, which necessitates the adjustment of the head gauges such that the zero level is at the same elevation as the flume throat.

Cutthroat Flumes. The cutthroat flume was developed for use in flat-gradient channels, where a flume was needed to operate satisfactorily under both free (critical) flow and submerged flow conditions (see Grant, 1989).

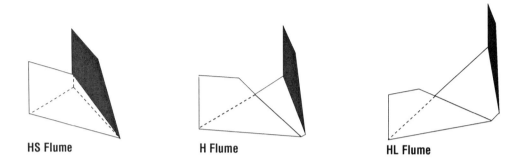

HS Flume   H Flume   HL Flume

Figure 3.14   HS, H, and HL flumes.

Figure 3.15   HS, H, and HL flume application.

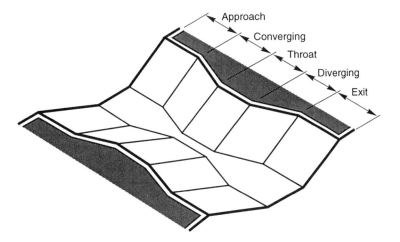

Figure 3.16 Trapezoidal flume.

Secondary Measuring Devices. As discussed above, the flow rate or discharge through a weir or flume is typically a function of the liquid level in or near the primary measuring device. A secondary measuring device (or open-channel flow meter) is used in conjunction with a primary measuring device to measure the rate of liquid flow in an open channel. The secondary measuring device has two purposes:

- Measuring the liquid level in the primary measuring device, and
- Converting this liquid level to an appropriate flow rate.

This flow rate information can then be transmitted to another device such as a totalizer, automatic sampler, or recorder.

The most important task for secondary measurement devices is the accurate measurement of the liquid level in the flow stream. There are several methods of determining this liquid level, including the use of

- Floats,
- Electrical sensors,
- Ultrasonic sensors,
- Submerged pressure sensors, and
- Bubbler systems.

Each technique has specific applications to which it is best suited: inaccurate readings related to the nature of the device can result from using a specific technique in an inappropriate situation. Any of these devices can be used in a testing laboratory and produce excellent results. The accuracy of any of these methods depends on the actual stream measurement conditions. A good knowledge of field conditions is needed to make an accurate decision about the type of level-measurement system to use.

The conversion of the liquid level to useful flow rate information can also be accomplished through several methods. Early flow meters used mechanical cams and other optically read flow conversions. Today, flow meters typically use computer-generated tables or on-line devices that can perform instantaneous flow rate calculations.

AREA-VELOCITY METER. An additional method used to obtain a flow rate in an open channel is detection of the velocity of the flow stream. The area velocity method consists of measuring both the cross-sectional area of the flow stream at a certain point and the average velocity of the flow in that cross section. The flow rate is then calculated by multiplying the area of the flow by its average velocity. This calculation is often referred to as the continuity equation:

$$Q = A \times V \tag{3.1}$$

Where

Q = the flow rate,
A = the cross-sectional area of the flow, and
V = the average velocity of the flow stream.

In open channels, the area velocity method requires two separate measurements: one to determine the flow depth and the other to determine the mean velocity. The depth measurement is used to calculate the cross-sectional area of the flow based on the size and shape of the channel. The channel can be of any shape, such as round, U-shaped, rectangular, or trapezoidal. The area velocity method can also be applied to a channel of nonuniform shape as long as the relationship between level and area can be determined. If the channel geometry is known, the flow rate can be calculated.

The area velocity method has several advantages over weirs and flumes. The first advantage is that, in addition to measuring flow under free-flow conditions, this method can also be used to measure flow under submerged, full-pipe, surcharged, and reverse-flow conditions. These conditions can occur in a variety of circumstances, such as when undersized sewers, inflow and infiltration, or tidal effects are present. Weirs and flumes cannot be used to measure full-pipe, surcharged, or reverse flows.

The area velocity technique can be used in two ways. First, the depth and velocity can be measured manually and used to determine the area of flow and the flow rate at a particular time. Second, an area velocity flow meter can be used to measure the liquid level and velocity in the channel and automatically calculate the flow rate.

Propeller-type meters have typically been used for stream measurement (Smoot, 1974, and Parr *et al.*, 1981). However, new advances have led to the use of electromagnetic or Doppler technology to provide greater accuracy in both short-term profiling and continuous-monitoring applications.

CLOSED-PIPE FLOW MEASUREMENT. Wastewater flow through completely filled pressure conduits is referred to as *closed-pipe* flow. Pressure conduits are typically used for fresh water lines or industrial process lines, and the flow through them is often measured by some type of device placed in the line. Common closed-channel, flow-measuring devices include Venturi meters, ultrasonic meters (both Doppler and transit time), flow nozzles, orifice meters, magnetic flow meters, and pitot tube flow meters.

The measurement of flow in closed conduits is typically more accurate than that in open channels because of the additional control over variations in flow not available with open-channel measurement. All of the measurement devices described in this section are designed to measure flow in pipes flowing full. These devices can generally be mounted in any orientation, but the best practice is to install them on an incline with upward flow. This precaution prevents the existence of both a partially full line and diffused air entrapment. The specific flow velocity curve will affect all measurement devices. As with all flow measurement devices, periodic calibration is recommended to ensure accuracy.

Propeller Meter. The propeller meter (Figure 3.17) operates on the principle that liquid flowing against the propeller pitch will cause the propeller to rotate at a speed proportional to flow rate. The meter is self-contained and requires no auxiliary energy or equipment other than a mechanical totalizer to obtain the flow reading (not flow rate). Equipment can be added to the meter to elicit flow rate readings, pace chemical feed equipment, and actuate telemetering equipment for remote readout.

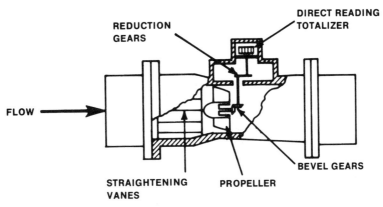

Figure 3.17   Propeller meter.

Accuracy for propeller meters is usually stated at ±2% of actual flow rate over a range of 7:1 peak-to-minimum flow ratio for small meters to 12:1 for large meters. Such meters should not be used either with liquids containing suspended solids or with those that deposit substances that can build up on exposed surfaces of the device and cause plugging. Slippage of the fluid on the propeller, friction in the mechanism, and fluid velocity sensing ability are all elements that can enter into meter inaccuracies.

Magnetic Flow Meter. The magnetic flow meter (Figure 3.18) has an insulating inner liner and operates on the principle developed by Michael Faraday's law of electromagnetic induction: any conductor, such as a column of conductive liquid passing through the lines of force of a fixed magnetic field, will generate an electromotive force (dc voltage) directly proportional to the rate at which the conductor is moving through the field.

Although the fluid velocity profile affects all types of flow-measuring equipment, the magnetic flow meter is the least affected by such variations within the profile.

To provide scouring, this meter must be sized for a minimum velocity of 1.5 to 2.4 m/s (5 to 8 ft/sec) when used on sludges. The system accuracy is usually stated at ±1% of the meter scale for a velocity range of 0.9 to 9.0 m/s (3 to 30 ft/sec). At less than 0.9 m/s, the system accuracy is ±2% of scale or lower. The presence of entrained gases will affect the flow readings. Grease or other substances that deposit on or coat the electrodes will also prevent proper performance. Heat or sonic cleaners can be used to prevent degradation of meter accuracy.

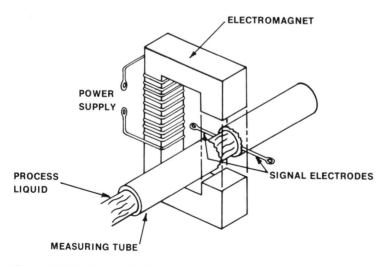

Figure 3.18   Magnetic flow meter.

For most installations, the error caused by piping configuration is less than ±0.5% of the meter accuracy. Therefore, it is not necessary to consider piping arrangements when calibrating.

Flow Tubes. The term *flow tube* (Figure 3.19) refers generically to devices installed in pressurized pipelines. A flow tube may be as simple as a straight piece of pipe in a pressure system or as sophisticated as a Venturi tube designed to relate flow rate to pressure drop across a constriction in the pipe. There are varying amounts of published data describing flow tube performance. Flow tubes must be calibrated individually if their coefficients are not repeatable (ASTM, 1976).

The accuracy of flow tubes is typically ±1% of actual flow rate. The pressure connections, when used in wastewater and sludge applications, should be equipped with pressure sensors that pick up the differential pressure while containing the line fluid within the line. Upstream piping configurations will significantly affect the accuracy of measurement.

Venturi Tube. The Venturi tube (Figure 3.20) operates on the principle that a fluid flowing through a meter section having a convergence and constriction of known shape and area will cause a pressure drop at the constriction area. This pressure difference is proportional to the square of the flow. The difference in pressure between the inlet and the throat pressure is proportional to the square of the flow. The accuracy of the Venturi generally is ±0.75% of flow rate. However, overall system accuracy will vary.

The pressure connections, when used for wastewater and sludge applications, should be equipped with water-purge systems or pressure sensors that pick up the differential pressure while containing the line fluid within the line.

The Venturi is sensitive to upstream pipe configurations, and the range of the Venturi is usually limited by the transducing device. On clean water, a sensitive transducer can measure a differential as low as 13 mm (0.5 in.) within ±1% of rate.

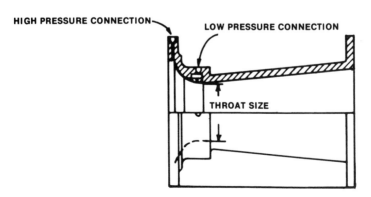

Figure 3.19   Flow tube.

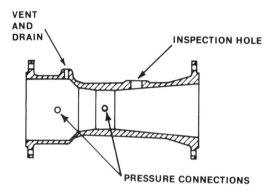

Figure 3.20   Venturi tube.

Ultrasonic Flow Meter. The ultrasonic flow meter is a noncontacting, flow-measuring system that can be installed in pipelines carrying liquids. It uses the Doppler principle stating that a change in the frequency of sound waves occurs as a result of a disturbance in the media transmitting the waves. The apparent change in frequency will be proportional to the velocity of the moving particles in the liquid.

The ultrasonic flow meter system comprises two interrelated components. The first is the sensor consisting of a precalibrated, typically fused, epoxy-coated flow tube. The sensor contains a pair of ultrasonic transducers mounted diagonally to the exterior of the tube section. The second part of the system is the transmitter or electronics package. These ultrasonic transducers are capable of sending and receiving ultrasonic pressure pulses. During operation, pulses are alternately transmitted, first against and then with the direction of flow, through the fluid. The pulse transmit time (upstream and downstream) can be expressed in terms of two frequencies because frequency is the reciprocal of period. The average fluid velocity is proportional to the difference between these two frequencies. The meter calculates the frequency differences, and its output is proportional to the difference between them.

The ultrasonic pulses can sometimes be blocked or attenuated by suspended solids or gas bubbles in the fluid. This blockage has presented problems in the past in applying the ultrasonic flow meter technology. The manufacturers of current meters claim that this problem has been overcome in two ways: first, through the use of high-strength pulse trains, high-gain preamplifiers that are sensitive even to severely attenuated signals; and second, through the use of specialized circuitry that rejects pulses that do not conform to a particular pattern. Ultrasonic flow meters, like most accurate flow meters, require 10 pipe diameters upstream and 5 diameters downstream of straight piping run because of sensitivity to pipe configuration patterns.

Accuracy at velocities of 0.3 m/s (1 ft/sec) or greater is specified to be within ±1% of reading for Newtonian liquids having Reynolds numbers higher than 100 000. Most wastewater applications for flow measurement fall in this range.

Variable-Area Meters. The variable-area meter, or rotameter (Figure 3.21), operates as a vertical flow tube with the float position being a function of viscous drag—that is, differential head or pressure. The constriction in the flow tube line is fixed, and flow is indicated by the pressure drop. In a vertical flow tube, the annular area between the float and flow tube constrictions changes in proportion to the flow through the annular space. As the flow increases, the height of the float above the constriction increases; conversely, as the flow decreases, the float drops.

The variable-area meter consists of a plummet or float and an upright tapered tube. The plummet is lifted to a state of equilibrium represented by the downward force of the plummet and the upward force of the flowing fluid (which must be free of solids) as it passes through the annular orifice around the plummet. The reading scale, attached to the tube, is essentially linear because of the constant differential pressure. There is always the possibility of mechanical resonance, which causes plummet oscillation and subsequent difficulties in the readings. Damping devices can reduce this problem. Some installations use a magnetic float that drives an external magnetic follower to indicate the flow rate (see Figure 3.21).

Rotameter accuracy is typically ±2% of maximum scale. Equipment costs generally limit the rotameter to pipe sizes of less than 60 mm (2.5 in.). These values change for fluids of different specific gravities and viscosities. Also, if different fluid densities are encountered, differently weighted or tailored floats will be required to achieve proper accuracy.

Orifice and Segmented Meters. The concentric orifice meter is a thin plate, placed into a pipe line, with a sharp, square-edged orifice that generates a pressure differential across the plate. The use of upstream and downstream pressure taps allows the indirect measurement of flow (Figure 3.22).

In some cases, concentric orifice meters can trap solids on the upstream side, resulting in inaccurate flow data. To solve this problem, segmented

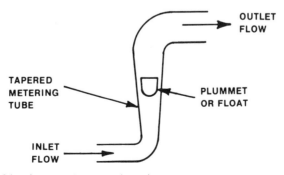

Figure 3.21   Area meter or rotameter.

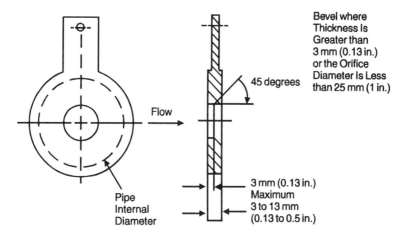

Figure 3.22   Concentric orifice plate.

orifice meters have been developed to allow the passage of the solids under the plate. The relative accuracy of these meters is approximately ±4% (Spink, 1967).

DENSITY MEASUREMENT

Some process control variables, such as mixed-liquor suspended solids in the activated-sludge process, depend on the densities of the treated wastewater and the removed solids. Methods of determining density are discussed in the following paragraphs.

GAMMA RADIATION. Gamma radiation detectors are composed of a source consisting of radioactive material that emits energy in the gamma region and a detector of this energy. The radiation source is placed opposite the detector inside a pipe. Both water and solids act as absorbers of gamma radiation. When the pipe contains only water, a radiation base level is detected. As solids replace the water in the area of detection, the detected radiation level decreases. This relationship allows a determination of the density (amount of solids present) in the flow stream: the lower the detected radiation, the higher the proportion of solids in the flow stream.

Periodic recalibration of the detector is necessary because the radioactive source gradually decays. Additionally, the presence of entrained gases causes erroneous readings in the gamma detector. Federal certification is required for individuals who handle radioactive material. These factors combined have limited the use of this density detector.

ULTRASONIC SENSOR. Ultrasonic density detectors consist of an ultrasonic transmitter, a receiver, and an electronic control unit. The control unit generates an electric signal that is converted to an ultrasonic signal. This signal is directed across a pipe section from the transmitter. The signal passes through the sludge, where it is detected by the receiver. This signal is then re-converted to an electronic signal received by the controller. Finally, the signal is sent to a relay that activates a control timer. Density ranges in which this device can be applied are 1% to 10%. Repeatability of the equipment is ±0.5%.

NONLIQUID SAMPLING

GASES. As in liquid sampling, gases may be collected as grab or composite samples (Figure 3.23). Grab sampling is normally performed using a glass or metal evacuated chamber. After it is filled with gas, the chamber is taken directly to the laboratory for analysis.

A variety of gas monitors is available for field analysis or composite gas collection. To determine carbon dioxide, oxygen, nitrogen, and methane, laboratory analysis using a gas partitioner is a standard procedure. New field equipment can determine both hydrogen sulfide and the above-mentioned gases. When a more sophisticated laboratory service or the detection of a specific gas is required, the operator should consult with laboratory personnel and consider purchasing a small, battery-operated air-metering pump. When combined with the use of small columns packed with synthetic adsorbents, these pumps can collect samples for a wide variety of gases.

SHUNT METER FOR GAS FLOW. The shunt meter (Figure 3.24) operates on the same principle as the flow tube: it measures a pressure drop across

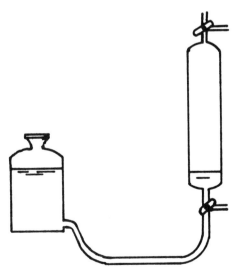

Figure 3.23 Displacement gas collectors are the most suitable containers when the gas source is some distance from the analyzing apparatus.

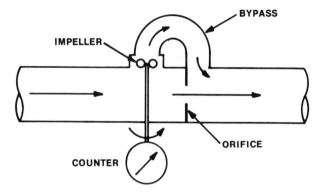

Figure 3.24   Shunt meter.

a metering orifice. The differential pressure across the in-line orifice will cause partial flow around the shunt. When properly dampened in a damping fluid, the impeller at the shunt orifice will rotate at a fixed ratio to the flow. Through the application of proper gear changes, this rotation can provide a direct readout of mainline flow rates.

In applications involving steam, the meter can be damaged if it is installed downstream of a valve that might trap condensate. Being a liquid, the condensate can strike and distort the impeller.

The accuracy of the shunt meter is typically ±2% of the actual flow rate for meters as large as 100 mm (4 in.) in diameter. Larger meters are typically not as accurate.

POSITIVE DISPLACEMENT DIAPHRAGM METER FOR GAS FLOW. The positive displacement diaphragm meter (Figure 3.25) operates through the alternate filling and discharging of a definite volume of gas from either side of the stroking diaphragm. As in home gas measurement, the mechanical motion required to turn a direct-reading register is imparted to a

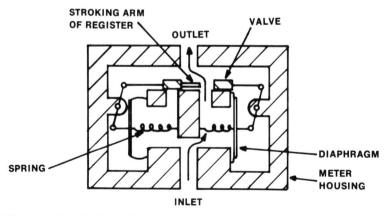

Figure 3.25   Positive displacement diaphragm meter.

lever and pawl arrangement. This low-pressure, wide-range device is used to measure digester gas. Accuracy is usually ±1% of actual rate.

SLUDGES. Equipment used to sample and analyze sludge (U.S. EPA, 1978) varies widely with the kind of facility and method of sludge handling. Sample collection will vary most with the moisture content of the sludge.

Automatic wastewater sampling equipment has been used for slurry streams containing as much as 35% solids. This procedure is not recommended, however, because the equipment is unreliable and the sample is typically unrepresentative. There are positive displacement sample collectors designed for sampling slurries, but so far these have found little use in wastewater treatment plant operations. Equipment such as displacement samplers should have increasing application in wastewater treatment systems, both for process control and for monitoring sludge quality for eventual disposal.

Where grab samples are desired in highly liquid situations, the Coliwasa sampler can be used. Where the sludge has a higher solids content, such as after dewatering or drying, the method of multiple partitioning of large representative samples is recommended.

REFERENCES

American Public Health Association (1995) *Standard Methods for the Examination of Water and Wastewater.* 19th Ed., Washington, D.C.

ASTM (1976) *1976 Annual Book of ASTM Standards.* Philadelphia, Pa.

Developing Water Quality Standards (1974) *Environ. Sci. Technol.*, **8**, 786.

Grant, D.M. (1989) *Isco Open Channel Flow Measurement Handbook.* 3rd Ed., Isco Environ. Div., Lincoln, Neb.

Harris, D.J., and Keffer, W.J. (1974) Wastewater Sampling Methodologies and Flow Measurement Techniques. EPA 907/9-74-005, U.S. EPA, Region VII, Kansas City, Mo.

Junk, G.S., *et al.* (1974) Contamination of Water by Synthetic Polymer Tubes. *Environ. Sci. Technol.*, **8**, 13.

Kirkpatrick, G.A., and Shelley, P.E. (1975) *Sewer Flow Measurement—A State-of-the-Art Assessment.* EPA-600/2-75-027, Environmental Protection Series, U.S. EPA, 1975.

Parr, A.D., *et al.* (1981) Point-Velocity Discharge Measurement Method for Sewers. *J. Water Pollut. Control Fed.*, **53**, 1.

Rabosky, J.G., and Koraldo, D.L. (1973) Gauging and Sampling Industrial Wastewaters. *Chem. Eng.*, **80**, 111.

Shelley, P.E. (1977) *Sampling of Water and Wastewater.* EPA-600/4-77-039, U.S. EPA, Washington, D.C.

Shelley, P.E., and Kirkpatrick, G.S. (1973) *An Assessment of Automatic Sewer Flow Samplers.* Hydrospace-Challenger, Inc., EPA-R2-73-261, U.S. EPA, Office Res. Monit., Washington, D.C.

Smoot, G.F. (1974) *A Review of Velocity Measuring Devices.* Geol. Surv. Open File Rep., U.S. Dep. Interior, Washington, D.C.

Spink, L.K. (1967) *Principles and Practice of Flow Meter Engineering.* The Foxboro Company, Foxboro, Mass.

U.S. Environmental Protection Agency (1978) *Sludge Handling and Conditioning.* EPA-430/9-78-002, Office of Water Programs, Washington, D.C.

Chapter 4
Quality Assurance
and Quality Control

*I*NTRODUCTION

Quality assurance (QA) is a system devised to ensure the quality of analytical data and the integrity of decisions that are made based on that data. The QA plan consists of programs for process control and employee training, a policy of continuous improvement, and a well-planned monitoring program designed to prevent poor data quality. There are two separate but related aspects of quality assurance: quality assessment and quality control. Quality assessment is the continuous statistical evaluation of analytical data to determine whether quality control (QC) activities are effective and the objectives of the QA program have been achieved. Quality assessment involves measuring both the validity and comparability of the data obtained. The QC, or auditing, program makes use of a variety of blanks, duplicate measurements, and spiked samples to gauge the validity of analytical results.

Although every aspect of the sampling program must be carefully planned, the QA program in particular requires a great deal of forethought. Not only do the number and types of field QC samples need to be considered, but attention must be given to the smallest of details, such as the selection of sample containers.

Sampling equipment needs may vary depending on the site, the analytes of interest, and the test methods employed. It is crucial that field personnel be trained to properly operate sampling equipment, follow sampling standard operating procedures (SOPs), use approved test methods, and maintain accurate field records. Training should also address required container types, preservatives, and recommended holding times for the analytes of interest. This information can be found in Chapter 5 of this manual and in 40 CFR Part 136.3, 7-1-90 edition, Table II, and can be distributed to personnel for quick reference in the field. Employees involved in sampling operations should have an understanding of the analytical method requirements as they apply to sample collection. For example, the test procedure for volatile organics requires sample vials to be filled completely, with no head space, while the method for total filterable residue recommends a head space sufficient to allow adequate mixing. Equipment cleaning practices should also be discussed during training. Thorough cleaning of equipment and containers is as important to data quality as proper sample handling and preservation procedures.

SOURCES OF ERROR

ANALYTICAL INTERFERENCES. Analytical interferences are often sources of error in environmental applications. An interference can be defined as a physical or chemical property of a sample that causes error in the measurement process (Lewis, 1988). There are two recognized types of interferences: additive and multiplicative. Additive interferences are caused by some constituent of the sample that produces a signal similar to that of the analyte of interest, thereby yielding a reading much higher than what is attributable to the analyte concentration. Multiplicative interferences are caused by a component of the sample that will cause the analyte signal to increase or decrease without generating its own signal (Lewis, 1988). Matrix effects are typically the most common multiplicative interferences; however, sample contaminants may also cause interference because of absorptive loss of the analyte. Refer to Chapter 5 for additional information concerning sample handling.

SAMPLE CONTAMINATION. Contamination of environment samples is commonplace and can originate from sources such as carryover in the sample containers or sampling lines; cross-contamination during shipment; improper sample handling, transportation, or storage; or even ambient air. In fact, the actual sampling devices may contribute to cross-contamination of samples. Thus, the more sensitive analytical methods may stipulate the use of specialized sampling materials. Priority pollutant scans, for example, require that automatic samplers be equipped with polytetrafluoroethylene or polytetrafluoroethylene-lined tubing and glass collection containers.

Dirty sampling equipment, or devices that have not been properly decontaminated, may also corrupt a sample because of carryover from a previous sample. The selection of cleaning procedures for sampling equipment and containers should include a consideration of all of the parameters being analyzed. When several different tests are performed on a single sample or subsamples from one container, a procedure that minimizes contamination of some analytes may actually be a source of contamination for others (Lewis, 1988). Ensure that the parameters of interest are not components of the detergents or acids being used to decontaminate sampling equipment. Although no single cleaning method is suitable to all parameters and types of analysis, a mild soap solution followed by a rinse of distilled or deionized water will typically suffice when wastewater and sludge samples are to be tested for conventional parameters. Use only soap that is non-phosphate-based and has been tested for toxic effects by a certified laboratory. Some analyses require special cleaning procedures ranging from acid- or solvent-washing to sterilization. Before initiating any cleaning process, refer to both the equipment manufacturer's instructions and the analytical methods involved for precautions and recommended procedures.

Sample preparation in the field presents another potential route of contamination, but field manipulation is often unavoidable. Accordingly, all sample handling and subsampling activities should be carefully planned, documented, and kept to a minimum. Errors can result from contamination that is caused by material transfers, filtration, physical measurements, aliquot preparation, or spiking (Keith, 1991). In one case, during the collection of samples for soluble metals analysis, filtration devices were rinsed with 5% nitric acid to minimize trace metal contamination. Nitrate contamination as high as 3.5 mg/L was found after filtering of aqueous samples, even when the nitric acid rinse was followed by thorough rinsing with deionized water (Lewis, 1988).

The sample preservative may also introduce bias if it contains significant concentrations of the analyte of interest. To avoid this situation, purchase chemicals that are American Chemical Society grade or better, and check the product specifications list for possible contaminants and their expected concentrations. Because it is also likely that chemical preservatives will become contaminated after continued use in the field, it is advisable that sample collection bottles be prepared in the laboratory and transported to the sampling site. When this is impossible or impractical, a small portion of the preservative may be transferred to a clean, dry, properly labeled reagent bottle and dispensed to the sample collection bottles in the field. Any remaining preservative should be disposed of in accordance with an appropriate hazardous waste disposal policy. Unused aliquots of the preservative should never be returned to the original chemical container.

SAMPLING QUALITY CONTROL

Field QC samples are used to identify and control errors resulting from interferences and contamination. The number and types of QC samples collected will depend on the individual sampling program objectives. It is important that field QC samples be handled in exactly the same manner as any other environmental sample. Identical sampling devices, sampling protocol, storage containers, shipping procedures, and preservation techniques should always be used.

Just as there are numerous types of laboratory QC samples, there are several types of field QC samples, including blanks and samples used to assess precision and accuracy. This chapter, however, will address only laboratory blanks and controls as they relate to sampling operations.

BLANKS. Blanks can be either natural samples or sample matrices that contain negligible amounts of the substance of interest and are representative of the environment and media being sampled. Most blanks are aqueous, but those based on analytical matrices other than water vary widely in composition and consistency. Blanks are used to gauge the results of specific sampling or test procedures.

Blanks are used to reduce error by compensating for interferences and identifying contaminants that may not be quite as obvious as those mentioned previously. According to Keith (1991), blanks should be devised to detect and measure contaminants wherever the possibility exists for introducing extraneous material to collection, treatment, or analytical procedures. Sample blanks have both assessment and control functions. When used to evaluate sampling operations, blanks define the limitations of analytical measurements, identify points of contamination, and, in some cases, can be used to estimate the concentration of the contaminant. Blank data can also be used for adjusting sample test results to compensate for background contamination. These adjustments should be made cautiously, however. Taking the average of multiple blank measurements helps ensure a stable measurement system (Lewis, 1988). Most laboratories, however, will not report "adjusted" data. Data from blanks can be plotted on a control chart versus time and, in this way, used to initiate corrective action when contamination is indicated.

The use of blank data as a control mechanism has two significant drawbacks: it requires a relatively long period of time, and there typically is not ample opportunity to resample when blanks show contamination. Consequently, blank data are used more often for assessment than control.

Depending on whether or not they are used properly, blanks can either be helpful for assessment or they can lead to increasingly variable and misleading data. Blanks can be used to detect additive interferences, but they cannot effectively identify interferences such as dilution or adsorption; nor can they be used to identify noncontaminant sources of error such as analyte losses resulting from volatilization or decomposition (Lewis, 1988).

Two types of blanks used for environmental sampling applications are trip and field blanks. While trip blanks are quite specific in nature, the term *field blank* encompasses a broader classification that includes matrix or matched-matrix blanks, equipment blanks, preservative blanks, and preparation blanks.

Volatile Organic Compound Trip Blanks. Trip blanks, or transport blanks, are samples of analyte-free media taken from the laboratory to the sampling site and returned to the laboratory unopened. They are used mainly to detect sources of contamination that could potentially affect the analyte values of the actual sample in either a quantitative or qualitative fashion (NJDEPE, 1992). The trip blank is used to estimate cross-contamination from the container and preservative during shipment and storage of samples. The U.S. Environmental Protection Agency (U.S. EPA) has shown that aqueous samples can be contaminated during transportation and storage by diffusion of volatile organic compounds (VOCs) through the septum and around the screw cap of sample vials. Trip blanks are now used exclusively for aqueous VOC samples (Keith, 1991).

A trip blank is prepared by filling a clean 40-mL vial with distilled or deionized water and sealing the cap so that there is no head space. The blank is never opened in the field. It travels to the site with the empty sample bottles and back from the site with the collected samples in an effort to simulate sample handling conditions. Volatile organic compound trip blanks must be returned to the lab, stored, and later analyzed with the same set of bottles they accompanied to the field. One trip blank should be taken per day for each type of sample (NJDEPE, 1992). If contamination of a trip blank is indicated, a check of both the quality of the water used and the bottle-cleaning methods may also be warranted.

Field Blanks. The term *field blank* is used to describe any type of blank used to assess field contamination during a sampling event. Field blanks are used as controls on sample equipment handling, preparation, storage, and shipment, beyond that which is intended for trip blanks (NJDEPE, 1992). Like the trip blank, the field blank represents the effects of container preparation and transportation on sample quality. Unlike the trip blank, however, the field blank is used throughout the entire sample collection and handling process. Thus, the blank sample is exposed to the same potential sources of contamination as the actual environmental sample (Lewis, 1988).

Capped and cleaned bottles containing distilled or deionized water are taken to the sampling location where they are opened to the atmosphere, transferred from one vessel to another, or exposed to some other aspect of the sampling environment. Although it may be difficult or impossible to duplicate the exact process by which the actual sample was taken, the blank sample should be exposed to as many of the same elements of the process as possible (Lewis, 1988). Field blanks should be preserved in the same manner as samples and tested for the exact parameters as the associated samples. They should be collected at a rate of one blank per day for each collection apparatus.

There are several different types of field blanks. When designing the QA project plan, a consideration should be made of which types will be most useful to the sampling program. The frequency of analysis for each sample type outlined below is recommended elsewhere (NJDEPE, 1992). It is advisable to consult with applicable state agencies for their recommended frequencies, as their requirements may differ from those listed in this chapter.

MATCHED-MATRIX FIELD BLANK. With matched-matrix blanks, the most common types of field blanks, the blank matrix closely resembles that of the sample. Water blanks are widely used for aqueous sampling because they are easy to prepare, and deionized water is readily available.

It is more difficult to prepare matched-matrix field blanks for solids sampling. For soil-like material, reagent-grade sand (such as silica) can be used to obtain a similar sample matrix. If the analytes include organic compounds, baking the sand in a 300°C oven for several hours should remove volatile species. An exact matched-matrix field blank may be difficult or impossible to obtain when sampling solid or liquid wastes, sludges, or slurries. In this case, an approximation will usually identify most contaminant sources. Field blanks can be prepared for these situations by using uncontaminated soil or reagent sand or preparing aqueous or solvent mixtures of soil or sand (Lewis, 1988).

Matched-matrix field blanks for both aqueous and nonaqueous matrices should be taken at a rate of one blank for every 10 samples collected.

EQUIPMENT BLANKS ("DECON" OR FIELD RINSEATE BLANKS). An equipment blank is used to check environmental samples for cross-contamination and verify the effectiveness of equipment cleaning methods. To prepare an equipment blank, the decontaminated sampling device is rinsed with deionized water or an appropriate solvent. The rinseate is then collected and preserved in the same manner as the samples taken at that site. An analysis of the rinse solution will indicate the types of contaminants that may have resulted from contact with the sampling equipment (Lewis, 1988). As a rule, one equipment blank should be evaluated per day for each collection device.

PRESERVATIVE BLANKS. A preservative blank, as the name implies, is used to determine whether the chemical preservative has become contaminated. This blank is prepared by adding a volume of preservative to a bottle of deionized water equal to that added to the samples themselves. The preservative blank should be tested for the same parameters as the samples with which it is associated. For aqueous samples, the matched-matrix blank also serves as the sample preservative blank. For other matrices, preservative blanks may not be needed unless contamination is indicated by analysis of another preserved field blank. One preservative blank should be run with every batch collected.

PREPARATION BLANKS. A sample preparation blank is needed when blending, mixing, or subsampling of the original sample volume is done before sample analysis. This type of blank may be prepared using one of two methods (Black, 1988). The first, designed for an aqueous matrix, is similar to an equipment blank, in which distilled or deionized water is processed similar to the samples and using the same cleaned apparatus. The water is then collected, preserved, and later analyzed with its associated samples.

The second method is suitable for both aqueous and soil matrices and applies specifically to subsampling. Two identical sets of bottles are sent to the sampling site. Using clean devices, samples collected in one set of bottles are processed then poured into the empty set of bottles to be processed again. This method allows for evaluation of cleaning procedures and cross-contamination from both the processing equipment and the bottles themselves. One preparation blank should be collected for every 20 samples.

It is important that all sampling and field manipulations be conducted promptly to minimize sample exposure to airborne contaminants. According to Lewis (1988), sample contamination by lead, aluminum, and particulate sulfate are common in urban environments.

It is prudent to prioritize the types of blanks used in the sampling program, given the number of options and the expense associated with analytical services. One approach might be to collect a full series but analyze only one or two selected blanks. If contamination is indicated, the remaining blanks may be analyzed. Care should be taken, when using this approach, to avoid exceeding the prescribed holding times of the other blanks, which could render them invalid (Keith, 1991). A better strategy might be to reduce the likelihood of contamination at as many points in the sampling process as possible then collect those blanks that cover the broadest scope of sampling activities. If contamination is indicated, source-specific blanks can be added to the sampling regimen. The actual number and types of blanks needed depend on the intended use of data collected and the level of accuracy and precision needed to meet the study objectives.

ACCURACY AND PRECISION. The primary objective of any sampling program is to collect samples that are representative of both the media and the conditions being measured and to prevent poor data quality. The value of analytical data is determined by assessing both the accuracy and precision of the methods through which they were obtained.

Sampling Accuracy. Accuracy is a representation of the ability to achieve a true value or a measurement that is within an acceptable range of that value. Two sample types commonly used to evaluate the accuracy of environmental sampling are field spikes and blind standards. Both of these should be collected at a frequency of one per every 20 analyte samples, although a prudent course of action is to check with the regulating authority for frequency requirements during the preparation of the sampling plan.

FIELD SPIKE SAMPLES. A field spike sample is a selected natural sample to which a known amount of one or more analytes of interest is added at the time of collection. Field spike samples are used to measure analyte losses that occur from field handling, transportation, and storage of the samples. Analytical interferences are likely with some samples, such as sludge, because of the complexity of the matrix. A field spike, sometimes called a *matrix control,* can estimate the magnitude of those interferences and is often a good substitute for background samples. To some extent, a field spike can also be a measure of the competence and expertise of sampling personnel. Precise measurement techniques must be used during the preparation of these samples in the field; otherwise, erroneous data will complicate interpretation of the analytical results.

BLIND (KNOWN) STANDARDS. Blind standards are prepared chemical standards of known concentrations that have matrices similar to the batch of samples submitted to the laboratory by field personnel. Blind standards are used to evaluate the accuracy of the analytical data produced. The results of these samples are often used to prequalify a laboratory before its services are engaged on a contractual basis. The use of a certified standard, or laboratory control sample, is recommended. These standards can be manufactured using raw chemicals, purchased from a number of commercial vendors, obtained from state agencies, or be part of a past performance evaluation study.

Sampling Precision. Precision is a measure of the ability to reproduce consistent results and can be used to assess the condition of equipment and sampling techniques of field personnel. To allow the evaluation of sampling conditions, samples must be collected at random intervals and from two identical sets of field equipment.

There are two types of samples commonly used to check environmental sampling and analytical precision: duplicates and field split samples. These samples should be taken at a rate of one per analyte sampling event or one per every 20 samples of the same matrix collected. When preparing the sampling plan, check with the appropriate regulatory authority for QC sample frequency requirements.

DUPLICATES. Duplicate samples are used to evaluate intralaboratory performance by comparing analytical results of two aliquots from the same sample. Duplicates can also reflect the consistency of sample-handling processes in the field. Duplicate samples are collected simultaneously or in immediate sequence, homogenized, split from a larger volume, and submitted to the laboratory as blind samples. A duplicate aqueous sample should be prepared by alternately filling sample containers from the original sample container. To generate two equally representative samples, soils and sludges require homogenization. Always use clean stainless steel or polytetrafluoroethylene bowls and utensils when mixing these samples to avoid contamination. The sample should be divided in half and the containers filled by scooping sample material alternately from each half. Unless otherwise specified, volatile organic samples should be taken from discrete locations without mixing or compositing. This helps prevent the loss of volatile constituents and preserve the integrity of the volatile fraction of the sample (NJDEPE, 1992).

FIELD SPLIT SAMPLES. Split samples are used to evaluate interlaboratory performance. A field split sample is obtained using the same method outlined for a duplicate sample, where two representative subsamples are taken from the original volume. Both samples are tested for the same analytes by at least two different laboratories.

Field QC samples should be taken at regular intervals to signal when the system for collecting samples is lacking proper controls. Field QC samples are used to estimate sampling bias or calculate sampling precision (Keith, 1991). A summary of the various types of field QC samples, their purposes, and methods of collection can be found in Table 4.1.

*F*IELD *MEASUREMENTS AND INSTRUMENTATION*

Typically, it is more convenient to perform analyses in the field because sampling sites are generally remote and the samples require no preservation or transportation. However, some sample characteristics are dependent on conditions

Table 4.1   Summary of field quality control samples (Lewis, 1988, and U.S. EPA, 1990).

Sample type	Purpose	Collection	Documentation
Volatile organic component trip blank	Check contamination of sampling media from field to laboratory	Collect one sample (HPLC[a]-grade water) per each day of organic sampling	Assign separate sample number; record on chain-of-custody form
Field blank	Check cross-contamination during sample collection and shipment and in the laboratory	Collect for each group of samples of a similar matrix per each day of sampling; use HPLC-grade water (carbon-free) for organics; use metal-free (deionized or distilled) water for inorganics	Assign separate sample number; record on chain-of-custody form
Matched-matrix or matrix field blank	Detect and quantitate contamination introduced during sample collection, handling, storage, transport, preparation, and analysis	Made to simulate sample matrix and carried through the entire sample collection, handling, and analysis process; recommended frequency of analysis is once per 10 samples collected	Assign separate sample number; record on chain-of-custody form
Equipment field blank	Check field decontamination procedures	Check when sampling equipment is decontaminated and reused in the field or when a sample collection vessel (bailer or beaker) will be used; use blank water (HPLC-grade for organics, deionized or distilled for inorganics) to rinse equipment and pour this water into the sample containers; recommended frequency is one sample per day per device	Assign separate sample number; record on chain-of-custody form

Table 4.1 Summary of field quality control samples (Lewis, 1988, and U.S. EPA, 1990) (continued).

Sample type	Purpose	Collection	Documentation
Preservative field blank	Check for contamination of preservative	Add preservative to blank water (HPLC-grade for organics, deionized or distilled for inorganics) to rinse equipment and pour this water into the sample containers; analyze one sample per batch	Assign separate sample number; record on chain-of-custody form
Preparation field blank	Check sample handling and field decontamination procedures	Blank material (HPLC-grade water for organics, deionized or distilled water for inorganics, or reagent-grade sand for soil matrices) is processed and sub-sampled in field, then poured into empty set of bottles brought to site; analysis frequency is one sample for every 20 samples collected	Assign separate sample number; record on chain-of-custody form
Field spikes	Measure analyte loss resulting from handling, transportation, and storage	Add known concentration(s) of analyte(s) to deionized water or other solvent in the field; analysis frequency is one sample for every 20 samples collected	Assign separate sample number; record on chain-of-custody form
Blind standards	Check accuracy of laboratory	Purchase and prepare according to manufacturer's directions or prepare in laboratory; analysis frequency is one sample for every 20 samples collected	Assign separate sample number; record on chain-of-custody form

Table 4.1 Summary of field quality control samples (Lewis, 1988, and U.S. EPA, 1990) (continued).

Sample type	Purpose	Collection	Documentation
Duplicates and field-split samples	Check laboratory and field procedures	Collect from areas that are known or suspected to be contaminated; recommended analysis frequency is one sample per 20 samples collected	Assign two separate sample numbers; submit blind to the lab; record on chain-of-custody form
Calibration or field check standards	Measure instrument or method precision	Deionized water or other solvent containing a low but measurable concentration of the analyte of interest; should be the first sample analyzed; repeat every 20 samples	Data should be recorded in log book and analyzed statistically after every 20 data points
Calibration blanks and standards	Detect and quantify impurities in deionized water or other solvent; corresponds to zero analyte concentration	Consists of deionized water or other solvent used to dilute samples	All calibration data should be recorded in the field log book
Postcalibration or postbatch checks	Detect drift in calibration curve during analysis of a batch	One of the original calibration solutions or a known concentration falling within the linear portion of the calibration curve	Document in field log; data should be reviewed periodically; radical drifts are indicative of equipment malfunction

[a] HPLC = high performance liquid chromatography.

at the time of collection, and measurements must be carried out on site. Before any field testing is conducted, instrumentation and equipment must be calibrated or standardized.

EQUIPMENT CALIBRATION. Sampling equipment should be calibrated on a routine basis to minimize errors in flow rate, size discrimination, and temperature effects that could bias sample data. Standard operating procedures should be developed for the use, calibration, routine testing, and maintenance of all field equipment. Manufacturer's instructions can provide useful guidelines for developing these procedures. Written SOPs should establish testing and calibration intervals, a routine maintenance schedule, and a system of recordkeeping. It is also important to document the calibration standards and the environmental conditions, if applicable, required during calibration (U.S. EPA, 1982). Field equipment should be labeled with the date of the last calibration, the actual data obtained, and the calibration expiration date. Additionally, a complete record should be kept of all maintenance performed.

The process of instrument calibration is accompanied by its own set of QC samples. Those samples specific to the calibration of field instrumentation include calibration blanks, calibration standards, and postcalibration checks.

Calibration Blanks. A calibration blank is one of a series of standards used to calibrate an analytical instrument. The calibration blank, which represents a concentration of zero, can also be used to set the instrument response directly to zero before testing a sample. It is important that the power supply to the instrument not be interrupted but, rather, put in standby mode after calibration; otherwise, the calibration may be lost.

Calibration Standards. A calibration standard is based on a known concentration of a standard solution as determined by serial dilution. Serial dilution is a process by which an aliquot of a known solution having a relatively high concentration is diluted with more solvent, producing a larger volume of solution with a lower concentration. An aliquot of the resulting solution, in turn, is diluted in the same manner to produce a solution of even lower concentration. The number of standards required for calibration, and thus the number of dilutions that will have to be prepared, depends on the instrument to be calibrated. It is always a good practice to reference the manufacturer's instructions. As a rule of thumb, a minimum of four standards and a calibration blank is required for most concentration-based calibrations. The most dilute standard is typically measured first, followed by increasingly higher concentrations, until the calibration is complete.

Calibration Check Standards. The calibration check standard (CCS) is known by several different names, among them calibration control standard, field check standard, quality control calibration standard, and postcalibration or postbatch standard. Most often, this standard is an aqueous solution of the analyte of interest at a low but measurable concentration, brought to the field from the laboratory. The CCS can be performed using an aliquot of one of the original calibration solutions or any solution of known concentration that falls within the linear portion of the calibration curve. These checks are conducted after each batch has been tested to ensure that there has been no significant drift from the original calibration curve. When analyzed before the first sample, the standard deviation of the CCS is a measure of the instrument precision. The calibration check standard should be analyzed again after every 20 samples. Used in this manner, the standard deviation of the CCS is a measure of the method precision (Black, 1988). Postcalibration checks should be documented in the field log.

Many field instruments do not use concentration-based calibrations; therefore, calibration blanks, standards, and postbatch checks are sometimes not necessary for standardization. Some instruments require that calibration procedures be performed at regular intervals. These procedures are explained in detail in the owner's manual supplied with the equipment. Documentation of these calibrations should also include a label with the date of the last standardization, the actual data obtained, and the expiration date. In addition, written SOPs should list any standards or buffers, special environmental conditions needed, a routine maintenance schedule, and a record of all repairs and preventive maintenance performed. Table 4.2 lists pertinent information regarding standardization and calibration of some common field equipment.

EQUIPMENT AND MEASUREMENT INTERFERENCES. With most field equipment and test procedures, there are inherent difficulties that arise from interferences. Fortunately, most of these can be avoided as long as the operator is trained to recognize the interferences when they occur and, even more importantly, to take appropriate measures to prevent problems with equipment and instrumentation. During development of the sampling program, it is important to check with applicable regulatory agencies for acceptable calibration protocols.

Table 4.2   Quality assurance calibration procedures for field analysis and equipment (U.S. EPA, 1982).

Parameter	General	Daily	Quarterly
Automatic samplers	Enter the make, model, and serial and/or identification number for each sampler in a log book	Portable samplers: check volume setting using a class-A graduated cylinder; the same should be done for stationary samplers on a weekly basis	Check intake velocity versus head using a minimum of three samples and clock time setting versus actual time interval
Temperature			
Mercury thermometers	Enter the make, model, serial and/or identification number, and temperature range for each thermometer in a log book; standardize against an NIST[a] thermometer; readings should agree within 1°C, and data should be reported to the nearest 0.1°		Check at two temperatures against an NIST or equivalent thermometer; enter data in a log book; temperature readings should agree within 1°; if not, the thermometer should be replaced or recalibrated

Accuracy should be determined throughout the expected working range (0–50°C) both initially and annually thereafter; a minimum of three temperatures within the range should be used to verify accuracy |

Table 4.2 Quality assurance calibration procedures for field analysis and equipment (U.S. EPA, 1982) (continued).

Parameter	General	Daily	Quarterly
Thermistors, thermophones, and reversing thermometers	Enter the make, model, and serial and/or identification number for each instrument in field log; standardize against an NIST thermometer; readings should agree within 1°C, and data should be reported to the nearest 0.1°		Accuracy should be determined throughout the expected working range (0–50°C) both initially and annually thereafter; a minimum of three temperatures within the range should be used to verify accuracy
pH			
Electrode method	Enter the make, model, and serial and/or identification number for each meter in a log book; a record of all maintenance performed should also be logged	1. Calibrate the system against standard buffer solution of known pH value at the start of a sampling run 2. Periodically check the buffers during the sample run and record the data in the log sheet or book 3. Be on the alert for erratic meter response arising from weak batteries, cracked electrode, fouling, etc. 4. Check against the closest reference solution each time a violation is found 5. Rinse electrodes thoroughly between samples and after calibration	Take all meters to the laboratory for maintenance, calibration, and quality control checks
Dissolved oxygen			
Membrane electrode	Enter the make, model, and serial and/or identification number for each meter in a log book	1. Calibrate meter using manufacturer's instructions and Winkler-Azide method 2. Check membrane for air bubbles and holes; change membrane and potassium chloride if necessary 3. Check leads, switch contacts, and other parts for corrosion and shorts if meter remains unresponsive	Take all meters to the laboratory for maintenance, calibration, and quality control checks
Winkler-Azide method	Record data to nearest 0.1 mg/L	Triplicate analysis should be run to check the precision of the analyst; values should agree within ±0.2 mg/L	

Table 4.2 Quality assurance calibration procedures for field analysis and equipment (U.S. EPA, 1982) (continued).

Parameter	General	Daily	Quarterly
Conductivity	Enter the make, model, and serial and/or identification number for each meter in a log book; a record of all maintenance performed should also be logged	1. Standardize with potassium chloride standards having similar specific conductance values to those anticipated in the samples; calculate the cell constant using two different standards 2. Rinse cell after each sample to prevent carryover	1. Take all meters to lab for maintenance, calibration, and quality control checks 2. Check temperature compensation 3. Check date of last cell platinizing; replatinize if necessary 4. Analyze NIST or U.S. EPA reference standard and record actual versus observed readings in the log
Residual chlorine	Enter the make, model, and serial and/or identification number for each meter in a log book; a record of all maintenance performed should also be logged; report results to nearest 0.01 mg/L	Calibrate instrument (refer to the manufacturer's instructions for proper operation and calibration procedures)	Return instrument to lab for maintenance, calibration, and fresh reagents

[a] NIST = National Institute of Standards and Technology.

Automatic Samplers. With the advent of automatic samplers, human error in manual sampling activities has been virtually eliminated. Additionally, the use of automatic samplers has significantly reduced personnel costs associated with the performance of routine tasks and the collection of representative samples. Automatic samplers allow both for flow-proportional sampling and for more frequent sampling than what is practical using manual sampling processes.

Stationary automatic samplers should be calibrated according to manufacturer's specifications and checked routinely. Portable samplers, by nature, are more subject to wear and should be calibrated with each use. For additional information regarding automatic samplers, refer to Chapter 3.

Temperature Measurements. A mercury-filled thermometer, or equivalent, is suitable for most environmental temperature measurements. Because mercury is a hazardous compound, there is some concern regarding the use of mercury-filled thermometers in the field. Alternative types of thermometers are available. When a thermometer is to be used in the field, a plastic-coated thermometer with a metal case is recommended as a safety precaution. Some studies may require depth temperatures. These measurements can be taken with reversing thermometers, thermophones, or thermistors. Thermistors are by far the most convenient and accurate of the three. All thermometers and other temperature measurement devices must be calibrated annually against a National Institute of Standards and Technology traceable thermometer and documented in a permanent log book.

When using a thermometer, routinely check the mercury column for separations before taking the measurement. If there is a separation in the column, do not use the thermometer, but return it to the laboratory. Most column separations can be corrected by subjecting the thermometer to either heat or cold then bringing it back to ambient temperature. Separations in the contraction chamber can be rejoined quite simply in the field by inverting the thermometer and gently tapping it against the palm of your hand. Return the thermometer to its righted position, grip firmly, and hold it at a 90-deg angle to your body. From this position, swing the thermometer downward to your side in a 90- to 180-deg arc. Centrifugal force will cause the mercury to rejoin.

pH Meters. Calibration of pH meters is required to match the electrical response of the electrode to that of the instrument. The meter should be calibrated before each measurement when occasional pH measurements are made. However, when measurements are being made continually and the instrument response is stable, standardization is required only at 2-hour intervals. If pH values vary widely, standardize for each sample with a buffer having a pH range within 1 to 2 pH units of the sample (APHA, 1995). The meter should be standardized with the pH 7.0 reference buffer solution. For greater accuracy, both pH 4.0 and 10.0 reference buffers should then be tested to bracket the expected pH of environmental samples. Failure to obtain a reasonably correct pH value for these buffer solutions indicates a faulty electrode. A cracked glass electrode for instance, will often yield pH readings that are virtually the same for both standards (U.S. EPA, 1979). Glass electrodes are recommended because they are not affected by sample color or turbidity, colloidal matter, oxidants, reductants, or salinity, although a phenomenon called "sodium error" can be observed at pH values greater than 10 (APHA, 1995). Glass electrodes should not be allowed to become dry. When it is not in use, the electrode should be immersed in a 1:1 mixture of pH 7.0 buffer solution and deionized water or stored according to the manufacturer's recommendation. Electrode response may also be affected by oily materials coating the surface of the electrode, insufficient potassium chloride filling solution, or inadequate or improper maintenance. Often, normal response can be restored by cleaning the electrode using an appropriate procedure specified by the manufacturer.

Accurate readings are obtained slowly in weakly buffered samples and, therefore, should always be stirred during measurement. Stirring the sample also facilitates equilibrium between the electrodes and removal of carbon dioxide entrainment, as well as sample homogenization. When changing from buffered to unbuffered samples, the electrode should be rinsed with, or dipped several times into, the next test sample before the final reading is taken (U.S. EPA, 1979). Both glass and calomel electrodes should be rinsed with distilled or deionized water between each measurement and blotted with an absorbent tissue before the next sample measurement is taken.

Because of the asymmetric potential of the glass electrode, most pH meters are built with a slope adjustment that enables the analyst to correct for slight electrode errors observed during calibration with two or more different pH buffers (U.S. EPA, 1979). Adjustments to the slope must be made whenever electrodes are cleaned, changed, or refilled with fresh electrolyte. If the instrument does not have an automatic slope feature, a manual adjustment must be made that is appropriate to the ambient temperature, and all samples and buffer solutions should be measured at that temperature. A change in temperature affects the pH measurement for two reasons: it changes the electrode output, and it changes the pH of the sample. The first effect can be controlled through temperature compensation or by calibrating the instrument to the temperature of the samples. The second effect cannot be controlled because it is sample dependent. U.S. EPA recommends reporting both the sample pH and temperature at the time of analysis.

Dissolved Oxygen Meters. Dissolved oxygen (DO) meters must periodically be calibrated against a Winkler titration. This procedure, also called the *azide modification,* is described in APHA (1995). Because older units may require as long as 15 minutes to polarize after new electrolyte has been added, the instrument should be allowed to warm up while the Winkler titration is being conducted. Four 300-mL bottles should be filled with water of suitable quality. The titration should be set up in triplicate, and the values obtained should be averaged arithmetically. Any value that differs from the other two by more than 0.5 ppm should be disregarded. Place the DO probe in the fourth bottle, turn on the stirrer, and allow the probe to stabilize for 2 to 3 minutes. Adjust the DO meter reading to the average value obtained by the Winkler titration. Dissolved oxygen meters can also be air calibrated. This procedure is typically described by the instrument manufacturer in the owner's manual.

Erratic responses or off-scale readings may result from a clogged membrane or a membrane that is wrinkled or has entrapped air bubbles. Additionally, prolonged use of membrane electrodes in waters containing such gases as hydrogen sulfide (H_2S) tends to lower cell sensitivity (APHA, 1995). Changing the membrane and recalibrating the instrument at least once per week or whenever the membrane becomes wrinkled or clogged can prevent those interferences. If erratic responses persist, the oxygen sensor in the electrode may need to be replaced. Manufacturers typically provide troubleshooting guides to assist in pinpointing the source of the problem experienced.

Salinity also interferes with DO measurement because salt reduces the ability of water to hold oxygen in solution. When calibrating older DO meters using the Winkler method, the salinity of the calibration samples must be the same as that of the samples being measured. It is not necessary to know the salinity value, only that they are the same. Do not change the salinity setting after calibration. Most newer units have a salinity compensation feature.

High altitudes and atmospheric pressure also affect the DO reading; thus, the reading must be corrected for these effects. If atmospheric pressure is unknown, the altitude may be substituted. Refer to Table 4.3 for correction factors.

Conductivity Meters. Conductivity is the measure of a solution's ability to conduct electric current and is expressed in mhos (Mho). For the most accurate measurement, the cell of the conductivity meter should be standardized, and a cell constant should be established. First, rinse the cell by repeatedly immersing it in distilled or deionized water, then immerse the cell in the sample several times before obtaining a reading. If the meter is equipped with a magic eye, determine the maximum width of the shadow at least twice, once by approaching the endpoint from a low reading upward, and once from a high reading downward (U.S. EPA, 1979). For instruments that give measurement reading in mhos, the cell constant is calculated as follows:

$$L = k_1 + k_2/10^{-6} k_x \qquad (4.1)$$

Table 4.3   Altitude correction factor (APHA, 1995).

Atmospheric pressure, mm Hg[a]	Equivalent altitude, ft[b]	Correction factor
775	(−540)	1.02
760	0	1.00
745	542	0.98
730	1 094	0.96
714	1 688	0.94
699	2 274	0.92
684	2 864	0.90
669	3 466	0.88
654	4 082	0.86
638	4 756	0.84
623	5 403	0.82
608	6 065	0.80
593	6 744	0.78
578	7 440	0.76
562	8 204	0.74
547	8 939	0.72
532	9 694	0.70
517	10 472	0.68
502	11 273	0.66

[a] mm Hg × 0.133 3 = kPa.

[b] ft × 0.304 8 = m.

Where

L	=	cell constant;
k_1	=	conductivity, in micromhos per centimetre, of the potassium chloride solution at the temperature of measurement;
k_2	=	conductivity, in μMho/cm (mS/m), of the potassium chloride solution at the same temperature as the distilled water used to prepare the reference solution; and
k_x	=	measured conductance, in Mho (mS).

For instruments that read in the International System of Units (SI units), the following conversions can be applied:

$$1 \text{ mS/m} = 10 \text{ } \mu\text{Mho/cm} \qquad (4.2)$$

$$1 \text{ } \mu\text{Mho/cm} = 0.1 \text{ mS/m} \qquad (4.3)$$

Cell constants are prone to drift over time under ideal conditions; under adverse conditions, they are subject to change even more rapidly. Therefore, the cell constant should be established and documented on an annual basis.

Conductivity standards have a relatively short shelf life, so the expiration date should always be checked before using a standard solution. When primary standards are not readily available, the standard solutions recommended by, or purchased from, the manufacturer should be used. Standard concentrations should fall within the expected concentration ranges of the samples to be tested and near the lower instrument detection limit.

In natural water samples where the conductance is low, the usual reporting unit for conductance is micromhos (μMho). In samples having a conductance of less than 1 μMho, it is common to observe atmospheric effects caused by carbon dioxide and water forming carbonic acid ($CO_2 + H_2O \rightleftarrows H_2CO_3$). To minimize this effect when measuring highly purified water samples, take the reading while the water is running.

Temperature has a pronounced effect on conductance measurements. Because temperature variations between samples must be corrected for when reporting data, conductance data is reported at the standard temperature of 25°C. Temperature correction may be accomplished most simply by using electronic temperature compensation or by bringing all sample temperatures to 25°C, though the following mathematical formula can be used, if necessary.

$$\text{Reading}_{compensated} = \text{Reading}_{normal}/(P/4\%)(0.04T - 1) + 1 \qquad (4.4)$$

Where

P	=	dial setting in percent per 1°C, and
T	=	temperature in °C (U.S. EPA, 1979).

Instrument troubles are seldom encountered with conductivity meters because of the design simplicity (U.S. EPA, 1979). When problems do occur, they typically involve the cell. Cells should be examined frequently to ensure that the lead wires are properly spaced; the plates are clean, straight, and properly aligned; and the platinized coating is intact. Procedures for replatinization are often given in the equipment owner's manuals.

Chlorine Residual Measurements. Aqueous solutions of chlorine are not stable, particularly weak solutions. The concentration of chlorine will decrease rapidly on exposure to strong light or agitation. Because of this phenomenon, analysis should be conducted immediately after sampling to prevent loss of the analyte. There are a number of approved analytical procedures for chlorine residual measurement; selection of a method will largely depend on the types of samples to be analyzed. Chlorine residual may exist in a combined state with compounds such as ammonia, amines, and organic nitrogen and may be lost with the addition of certain reagents that alter the sample pH. Because the N,N-diethyl-p-phenylenediamine (DPD) colorimetric method is performed under neutral pH conditions, it is most often recommended for the analysis of wastewater. The DPD method is also perhaps the most practical for field measurements because of the convenience of the reagent pillows, the durability of the equipment, and the ease with which the equipment can be operated. As with each of the other methods, there are limitations associated with this procedure. High concentrations of monochloramine interfere with the free chlorine determination unless the reaction is stopped with arsenite or thioacetamide, and the DPD methods are subject to interference by oxidized forms of manganese unless compensated for by a blank (APHA, 1995). A sample blank must be used to compensate for color and turbidity in all colorimetric procedures.

DOCUMENTATION AND RECORDKEEPING

Perhaps the most important aspect of sampling is recordkeeping. Complete and accurate documentation of sampling events is imperative if data are to be defensible. Thus, a system for sample identification and documentation should be designed in accordance with the objectives of the sampling program. Records should consist minimally of labels, chain-of-custody procedures, and field log books. Computerization can be useful for establishing a sampling documentation system. In fact, many software programs can track inventories, provide for automatic sample login and label generation, and produce schedules for equipment calibration and maintenance.

Sampling and weather conditions should be recorded in the field log book. The log should include the sampling location and site description, weather conditions, sampling equipment and instrumentation used, date and time of sample collection, name of the field sampler, sample identification, results of all field measurements, and any other comments regarding the sample color, odor, appearance, or reactivity. A separate log book should be maintained listing all sampling equipment, including model and serial numbers and maintenance and calibration schedules. All samples must be labeled with a unique identification code, the date and time of collection, the sample collector's name, and the preservative used.

Documentation made on a chain-of-custody form should correlate with that in the log books and on the sample bottle labels. In addition to the items noted on the labels, the number of bottles, the analyses requested, and comments about the sample that might be helpful to the laboratory should be noted on the chain-of-custody form. The form must be signed by the sample collector and by every other person that handles, processes, or transports the samples after collection. A chain-of-custody form must accompany samples to the laboratory, where a member of the staff is required to sign for the samples on receipt. For additional information, refer to Chapter 6 on data collection and Chapter 7 on data preparation.

COMPLETENESS. The data are of little use unless they meet the criteria for completeness and comparability. Completeness is determined by the ratio of valid measurements obtained to the number of valid measurements needed to meet the data quality objective and should be documented for each measurement process. Data are considered valid if testing has been conducted in accordance with the analytical method and both the accuracy and precision of the measurements have reached predefined statistical levels of confidence in the resulting data, typically within 95% (U.S. EPA, 1989).

COMPARABILITY. Comparability is more of a qualitative measurement than completeness. Simply put, to facilitate comparison, analytical data must be calculated and reported in the standard units specified in the analytical method used, and they must be consistent with the applicable regulatory program for which the sampling is being conducted. Depending on the program objectives, it may be necessary to convert analytical data to different units based on another parameter such as flow. Often, at the analytical level, samples must be diluted to allow an accurate measurement of analyte concentrations. Before data are reported however, it is important that the test result, usually in milligrams per litre, be multiplied by the appropriate dilution factor. Whatever the conversion, all mathematical calculations should be documented in the field sampling and analysis logbooks. Records should be complete, accurate, and easy to read and understand (U.S. EPA, 1989).

SUMMARY

Analytical laboratories typically have extensive QA programs that include control charting for quality assessment and such QC measures as duplicate, split, and spiked samples; high- and low-range concentrations of known standard solutions; and method and reagent blanks. However, laboratory-oriented QC samples only identify errors that occur during sample preparation and analysis in the laboratory. To identify sources of contamination that are not attributable to laboratory error, field sampling activities must have their own QA/QC procedures. Quality assurance in sample collection should begin with proper employee training, and appropriate procedures should be implemented to minimize, to the extent possible, the likelihood of sample contamination.

There are many chances for error in sampling operations. The sampling program should be monitored for such common errors as improper sampling methodology, poor sample preservation, and insufficient mixing during compositing and testing (U.S. EPA, 1979). It is impossible to eliminate all sources of error and variability associated with collecting environmental samples; therefore, an effort must be made to keep all aspects of the sampling and analysis processes in a state of statistical control. This is accomplished through well-planned programs of quality control and statistical assessment.

REFERENCES

American Public Health Association (1995) *Standard Methods for the Examination of Water and Wastewater.* 19th Ed., Washington, D.C.

Black, S.C. (1988) Defining Control Sites and Blank Sample Needs. In *Principles of Environmental Sampling.* L.H. Keith (Ed.), Am. Chem. Soc., Washington, D.C.

Keith, L.H. (1991) *Environmental Sampling and Analysis: A Practical Guide.* Lewis Publishers, Inc., Chelsea, Mich.

Lewis, D.L. (1988) Assessing and Controlling Sample Contamination. In *Principles of Environmental Sampling.* L.H. Keith (Ed.), Am. Chem. Soc., Washington, D.C.

New Jersey Department of Environmental Protection and Energy (1992) *Field Sampling Procedures Manual.* Trenton, N.J.

U.S. Environmental Protection Agency (1979) *Handbook for Analytical Quality Control in Water and Wastewater Laboratories.* EPA-600/4-79-019, Cincinnati, Ohio.

U.S. Environmental Protection Agency (1982) *Handbook for Sampling and Sample Preservation of Water and Wastewater.* EPA-600/4-82-029, Cincinnati, Ohio.

U.S. Environmental Protection Agency (1989) *POTW Sludge Sampling and Analysis Guidance Document.* Washington, D.C.

U.S. Environmental Protection Agency (1990) *Sampler's Guide to the Contract Laboratory Program.* EPA-540/P-90-006, Washington, D.C.

SUGGESTED READINGS

American Chemical Society Committee on Environmental Improvement (1980) Guidelines for Data Acquisition and Data Quality Evaluation in Environmental Chemistry. *Anal. Chem.*, **52**, 2242.

Best, M.D., *et al.* (In press) *National Surface Water Survey, Eastern Lake Survey (Phase I-Synoptic Chemistry) Quality Assurance Report.* EPA-600/4-86-011, U.S. EPA, Washington, D.C.

Kerri, K.D., and Brady, J. (1989) *Industrial Waste Treatment: A Field Study Training Program.* Dep. Civ. Eng., Calif. State Univ., Sacramento.

Seanor, A.M., and Brannaka, L.K. (1984) *Influence of Sampling Techniques on Organic Water Quality Analysis.* Paper presented at Manage. Uncontrolled Hazardous Waste Sites Conf., Washington, D.C.

Stevenson, T.J. (1984) *Design of a Quality Assurance Program for the Assessment of Groundwater Contamination.* Paper presented at NWWA Conf. Groundwater Manage., Orlando, Fla.

Taylor, J.K. (1988) Defining the Accuracy, Precision and Confidence Limits of Sample Data. In *Principles of Environmental Sampling.* L.H. Keith (Ed.), Am. Chem. Soc., Washington, D.C.

Chapter 5
Sample Preservation, Handling, and Shipping

INTRODUCTION

Water and wastewater samples are collected for the purpose of obtaining, developing, and using representative data for a variety of needs. In addition to errors or changes that can be introduced by sampling equipment, methods, or techniques, potential biological, chemical, and physical changes can alter the composition of the sample. It is important to recognize the nature of these changes and know the steps that must be taken to minimize changes and prevent contamination resulting from external influences.

Complete preservation of samples, regardless of the source, is impossible. One's best efforts can achieve only a deceleration of the biological, chemical, and physical changes that inevitably will continue after sample collection. Although the complexity of these changes in a sample is beyond the scope of this manual, it is important to be aware of the general nature of the changes so that efforts can be made to control them.

NATURE OF CHANGES

The changes that occur in a sample can be classified into three general categories: physical, chemical, and biological. Physical changes within a sample occur when the sample environment is altered. This alteration may have an effect on either the biological or chemical nature of the sample. The sampling equipment or container used can significantly affect the data produced from a sample. Certain automatic sampling equipment can introduce contaminants from the outside. Vacuum-type samplers, which draw outside air for purging, can introduce both organic and inorganic substances to the sample, especially if the sampling atmosphere is near a pollutant-containing industrial area. Certain oil-lubricated systems that are not properly maintained can introduce contaminants that drastically alter the nature of the sample.

Certain cations are subject to adsorption on, or exchange with, cations in the walls of both glass and plastic containers and with the rubber or plastic hoses used in various sampling equipment. These cations include aluminum, cadmium, copper, chromium, iron, lead, manganese, zinc, and silver with certain glass containers. Sodium, boron, and silica can leach from glass containers. Plastic caps and liners (black Bakelite) and other rubber and plastic materials can contain cadmium, copper, iron, and zinc that can affect the sample's character. Residue from previous samples in a container is a common contaminant.

Temperatures change quickly and can affect both the chemical and biological characteristics of a sample. Elevated temperatures can solubilize precipitated material and increase the decomposition of organic materials. An increase in biological activity can be responsible for changes in nitrate, nitrite, or ammonia content, decreases in biochemical oxygen demand and phenol content, and other reduction reactions. Changes in pH also can occur rapidly as a result of changes in dissolved gas. For example, pH, alkalinity, or carbon dioxide shifts decrease the values of calcium and hardness.

Aeration of samples affects the dissolution of gases, which, in turn, can affect pH and other characteristics.

Zero head space is important for samples that are to be analyzed for volatile organics because these samples can volatilize in the container. In contrast, microbiological samples should have an air space at the top of the container for aeration and mixing.

Chemical changes in a sample's character primarily result from both physical and biological changes. Various metals can precipitate or solubilize, leach from container walls, or be reduced or oxidized. As a result of these changes, color, odor, or turbidity can decrease, increase, or change in quality. Photosynthetic activity from algae can change the carbon dioxide concentrations, altering the pH and, in turn, resulting in a change in various chemical equilibria.

Biological changes result from chemical and physical changes. As noted earlier, high temperatures increase biological activity, which results in changes in chemical composition and increased decomposition rates. Low temperatures, however, generally retard biological changes by decreasing biological activity and are considered a preservation technique.

Biological activity will result in the following:

- Changes in the dissolution of gases;
- Consumption of organic material;
- Production of cellular matter; and
- Formation of chemical byproducts such as carbon dioxide, ammonia, and water.

Certain constituents of a sample can undergo changes in their oxidation states. An example is the conversion of nitrite to nitrate. Soluble matter may be converted to organically bound material within the cellular structure; other cells may undergo lysis, releasing cellular material into solution.

It is apparent that the probability of changes occurring in a sample are high and that changes that occur in one area can multiply in effect in other areas. Physical, chemical, and biological changes are interrelated: a change in one will ultimately have an effect on all.

SAMPLING PROGRAM PLANNING

It is essential that any sampling program be preceded by a planning session at which the analytical requirements are considered (other details on planning for sampling are found in Chapter 2). This session should be attended by all persons to be involved in the sampling task, including laboratory analysts and sampling personnel. Most sampling plans will include on-site monitoring for such parameters as pH, temperature, and dissolved oxygen and will provide for preservation and preparation of samples to be delivered to the laboratory.

Planning must include consideration of sampling type, locations, equipment, and techniques; field monitoring to be performed; container types; volume requirements; preservation methods; and transportation requirements. The program must also be cognizant of quality assurance and quality control (QA/QC) requirements as described in Chapter 4.

When a sample, either grab or composite, is to be preserved with more than one reagent, it is necessary to split the sample into several subsamples. Also, if both total and dissolved parameters are to be analyzed, part of the sample must be filtered before any preservation reagents are added.

It is helpful to develop a flow diagram (Figure 5.1) of the sample distribution/preservation scheme. Table 5.1 lists the approved preservative for each parameter. This table is copied from 40 CFR Part 136.3 as revised July 1,

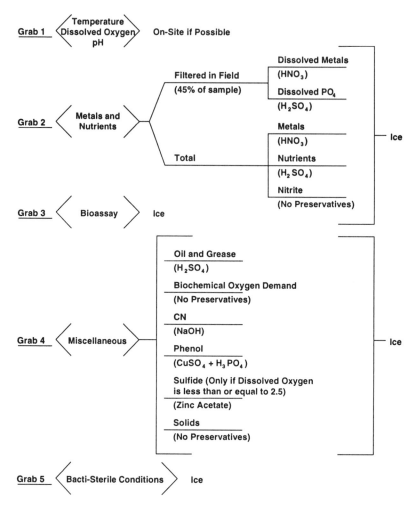

Figure 5.1 Flow diagram for a sample distribution/preservation scheme. Multiple samples will be required to obtain necessary sample volume; *Ice* indicates that all sample bottles are to be packed in ice; where acidification cannot be used because of shipping restrictions, the sample may be preserved initially by icing and then shipped immediately to the laboratory. Upon receipt in the laboratory, the sample must be acidified to a pH ≤ 2 with the appropriate acid.

1993. Using Figure 5.1 and Table 5.1, it is possible to develop the necessary checklists for field equipment, sample bottle, and preservative needs.

An additional aid for enhancing a sampling program is the use of detailed written standard sampling procedures. These help ensure that the appropriate sample is collected using the proper technique and equipment. These

Table 5.1 Required containers, preservation techniques, and holding times (U.S. EPA, 1993).

Parameter number/name	Container[a]	Preservation[b,c]	Maximum holding time[d]
Table 1A—Bacterial tests			
1–4. Coliform, fecal and total	P, G	Cool, 4°C, 0.008% $Na_2S_2O_3$[e]	6 hours
5. Fecal streptococci	P, G	Cool, 4°C, 0.008% $Na_2S_2O_3$[e]	6 hours
Table 1B—Inorganic tests			
1. Acidity	P, G	Cool, 4°C	14 days
2. Alkalinity	P, G	Cool, 4°C	14 days
4. Ammonia	P, G	Cool, 4°C, H_2SO_4 to pH<2	28 days
9. Biochemical oxygen demand	P, G	Cool, 4°C	48 hours
10. Boron	P (PFTE), or quartz	HNO_3 to pH<2	6 months
11. Bromide	P, G	None required	28 days
14. Biochemical oxygen demand, carbonaceous	P, G	Cool, 4°C	48 hours
15. Chemical oxygen demand	P, G	Cool, 4°C, H_2SO_4 to pH<2	28 days
16. Chloride	P, G	None required	28 days
17. Chlorine, total residual	P, G	None required	Analyze immediately
21. Color	P, G	Cool, 4°C	48 hours
23–24. Cyanide, total and amenable to chlorination	P, G	Cool, 4°C, NaOH to pH>12, 0.6 g ascorbic acid[e]	14 days[f]
25. Fluoride	P	None required	28 days
27. Hardness	P, G	HNO_3 to pH<2, H_2SO_4 to pH<2	6 months
28. Hydrogen ion (pH)	P, G	None required	Analyze immediately
31, 43. Kjeldahl and organic nitrogen	P, G	Cool, 4°C, H_2SO_4 to pH<2	28 days

Table 5.1 Required containers, preservation techniques, and holding times (U.S. EPA, 1993) (continued).

Parameter number/name	Container[a]	Preservation[b,c]	Maximum holding time[d]
Metals[g]			
18. Chromium VI	P, G	Cool, 4°C	24 hours
35. Mercury	P, G	HNO_3 to pH<2	28 days
3, 5–8, 12, 13, 19, 20, 22, 26, 29, 30, 32–34, 36, 37, 45, 47, 51, 52, 58–60, 62, 63, 70–72, 74, 75. Metals, except boron, chromium VI, and mercury			
38. Nitrate	P, G	Cool, 4°C	48 hours
39. Nitrate-nitrite	P, G	Cool, 4°C, H_2SO_4 to pH<2	28 days
40. Nitrite	P, G	Cool, 4°C	48 hours
41. Oil and grease	G	Cool to 4°C, HCl or H_2SO_4 to pH<2	28 days
42. Organic carbon	G	Cool to 4°C, HCl or H_2SO_4 or H_3PO_4 to pH<2	28 days
44. Orthophosphate	P, G	Filter immediately, cool, 4°C	48 hours
46. Oxygen, dissolved probe	G bottle and top	None required	Analyze immediately
47. Winkler	G bottle and top	Fix on site and store in dark	8 hours
48. Phenols	G only	Cool, 4°C, H_2SO_4 to pH<2	28 days
49. Phosphorus (elemental)	G	Cool, 4°C	48 hours
50. Phosphorus, total	P, G	Cool, 4°C, H_2SO_4 to pH<2	28 days
53. Residue, total	P, G	Cool, 4°C	7 days
54. Residue, filterable	P, G	Cool, 4°C	7 days
55. Residue, nonfilterable (TSS)	P, G	Cool, 4°C	7 days
56. Residue, settleable	P, G	Cool, 4°C	48 hours
57. Residue, volatile	P, G	Cool, 4°C	7 days
61. Silica	P (PFTE), or quartz	Cool, 4°C	28 days

Table 5.1 Required containers, preservation techniques, and holding times (U.S. EPA, 1993) (continued).

Parameter number/name	Container[a]	Preservation[b,c]	Maximum holding time[d]
64. Specific conductance	P, G	Cool, 4°C	28 days
65. Sulfate	P, G	Cool, 4°C	28 days
66. Sulfide	P, G	Cool, 4°C, add zinc acetate plus sodium hydroxide to pH>9	7 days
67. Sulfite	P, G	None required	Analyze immediately
68. Surfactants	P, G	Cool, 4°C	48 hours
69. Temperature	P, G	None required	Analyze
73. Turbidy	P, G	Cool, 4°C	48 hours
Table 1C—Organic tests[h]			
13, 18–20, 22, 24–28, 34–37, 39–43, 45–47, 56, 66, 88, 89, 92–95, 97. Purgeable halocarbon	G, Teflon™-lined septum	Cool, 4°C, 0.008% $Na_2S_2O_3$[e]	14 days
6, 57, 90. Purgeable aromatic hydrocarbons	G, Teflon™-lined septum	Cool, 4°C, 0.008% $Na_2S_2O_3$[e], HCl to pH2[i]	14 days
3, 4. Acrolein and acrylonitrile	G, Teflon™-lined septum	Cool, 4°C, 0.008% $Na_2S_2O_3$[e], adjust pH to 4–5[j]	14 days
23, 30, 44, 49, 53, 67, 70, 71, 83, 85, 96. Phenols[k]	G, Teflon™-lined cap	Cool, 4°C, 0.008% $Na_2S_2O_3$[e]	7 days until extraction; 40 days after extraction
7, 38. Benzidines[k]	G, Teflon™-lined cap	Cool, 4°C, 0.008% $Na_2S_2O_3$[e]	7 days until extraction[l]
14, 17, 48, 50–52. Phthalate esters[k]	G, Teflon™-lined cap	Cool, 4°C	7 days until extraction; 40 days after extraction
72–74. Nitrosamines[k,m]	G, Teflon™-lined cap	Cool, 4°C, store in dark, 0.008% $Na_2S_2O_3$[e]	7 days until extraction; 40 days after extraction

Table 5.1 Required containers, preservation techniques, and holding times (U.S. EPA, 1993) (continued).

Parameter number/name	Container[a]	Preservation[b,c]	Maximum holding time[d]
76–82. PCBs[k] acrylonitrile	G, Teflon™-lined cap	Cool, 4°C	7 days until extraction; 40 days after extraction
54, 55, 65, 69. Nitroaromatics and isophorone[k]	G, Teflon™-lined cap	Cool, 4°C, 0.008% $Na_2S_2O_3$[e], store in dark	7 days until extraction; 40 days after extraction
1, 2, 5, 8–12, 32, 33, 58, 59, 64, 68, 84, 86. Polynuclear aromatic hydrocarbons[k]	G, Teflon™-lined cap	Cool, 4°C, 0.008% $Na_2S_3O_3$[e], store in dark	7 days until extraction; 40 days after extraction
15, 16, 21, 31, 75. Haloethers[k]	G, Teflon™-lined cap	Cool, 4°C, 0.008% $Na_2S_2O_3$[e]	7 days until extraction; 40 days after extraction
29, 35–37, 60–63, 91. Chlorinated hydrocarbons[k]	G, Teflon™-lined cap	Cool, 4°C	7 days until extraction; 40 days after extraction
87. TCDD[k]	G, Teflon™-lined cap	Cool, 4°C, 0.008% $Na_2S_2O_3$[e]	7 days until extraction; 40 days after extraction
Table 1D—Pesticides tests			
1–70. Pesticides[k]	G, Teflon™-lined cap	Cool, 4°C, pH 5–9[n]	7 days until extraction; 40 days after extraction
Table 1E—Radiological tests			
1–5. Alpha, beta, and radium	P, G	HNO_3 to pH<2	6 months

[a] Polyethylene (P) or glass (G).

[b] Sample preservation should be performed immediately upon sample collection. For composite chemical samples each aliquot should be preserved at the time of collection. When use of an automated sampler makes it impossible to preserve each aliquot, then chemical samples may be preserved by maintaining at 4°C until compositing and sample splitting are completed.

[c] When any sample is to be shipped by common carrier or sent through the United States mails, it must comply with the Department of Transportation Hazardous Materials Regulations (49 CFR part 172). The person offering such material for transportation is responsible for ensuring such compliance. For the preservation requirements of Table II, the Office of Hazardous Materials, Materials Transportation Bureau, Department of Transportation, has determined that the Hazardous Materials Regulations do not apply to the following materials: hydrochloric acid (HCl) in water solutions at concentrations of 0.04% by weight or less (pH about 1.96 or greater); nitric acid (HNO_3) in water solutions at concentrations of 0.15% by weight or less (pH about 1.62 or greater); sulfuric acid (H_2SO_4) in water solutions at concentrations of 0.35% by weight or less (pH about 1.15 or

greater); and sodium hydroxide (NaOH) in water solutions at concentrations of 0.080% by weight or less (pH about 12.30 or less).

^d Samples should be analyzed as soon as possible after collection. The times listed are the maximum times that samples may be held before analysis and still be considered valid. Samples may be held for longer periods only if the permittee, or monitoring laboratory, has data on file to show that for the specific types of samples under study, the analytes are stable for the longer time, and has received a variance from the Regional Administrator under § 136.3(e). Some samples may not be stable for the maximum time period given in the table. A permittee, or monitoring laboratory, is obligated to hold the sample for a shorter time if knowledge exists to show that this is necessary to maintain sample stability. See § 136.3(e) for details. The term "analyze immediately" usually means within 15 minutes or less of sample collection.

^e Should only be used in the presence of residual chlorine.

^f Maximum holding time is 24 hours when sulfide is present. Optionally all samples may be tested with lead acetate paper before pH adjustments in order to determine if sulfide is present. If sulfide is present, it can be removed by the addition of cadmium nitrate powder until a negative spot test is obtained. The sample is filtered and then NaOH is added to pH 12.

^g Samples should be filtered immediately on-site before adding preservative for dissolved metals.

^h Guidance applies to samples to be analyzed by GC, LC, or GC/MS for specific compounds.

ⁱ Samples receiving no pH adjustment must be analyzed within seven days of sampling.

^j The pH adjustment is not required if acrolein will not be measured. Samples for acrolein receiving no pH adjustment must be analyzed within 3 days of sampling.

^k When the extractable analytes of concern fall within a single chemical category, the specified preservative and maximum holding times should be observed for optimum safeguard of sample integrity. When the analytes of concern fall within two or more chemical categories, the sample may be preserved by cooling to 4°C, reducing residual chlorine with 0.008% sodium thiosulfate, storing in the dark, and adjusting the pH to 6–9; samples preserved in this manner may be held for seven days before extraction and for 40 days after extraction. Exceptions to this optional preservation and holding time procedure are noted in footnote e (re the requirement for thiosulfate reduction of residual chlorine) and footnote l (re the analysis of benzidine).

^l Extracts may be stored up to 7 days before analysis if storage is conducted under an inert (oxidant-free) atmosphere.

^m For the analysis of diphenylnitrosamine, add 0.008% $Na_2S_2O_3$ and adjust pH to 7–10 with NaOH within 24 hours of sampling.

ⁿ The pH adjustment may be performed upon receipt at the laboratory and may be omitted if the samples are extracted within 72 hours of collection. For the analysis of aldrin, add 0.008% $Na_2S_2O_3$.

procedures should also detail the proper care and cleaning of equipment and containers and the provisions for using preservatives.

The sampling program should be reviewed periodically for QA/QC adherence (see Chapter 4). Any changes in laboratory analysis or sampling additions or deletions should trigger a review of the sampling program.

RESPONSIBILITIES OF SAMPLE HANDLING

The goal of most sampling efforts is to develop a database for decision making. Because the analytical results describing a system are only as accurate as the sample being analyzed, the key link in the analytical process is the proper collection and handling of the sample. Field personnel are responsible for the integrity of the sample from the time it is collected until it is in the custody of the laboratory. Poorly collected, damaged, or contaminated samples lead to incorrect data, which, in turn, can result in wastewater treatment plant malfunction, pollution of the environment, poor water quality, and a violation of treatment standards. In the case of an already malfunctioning plant, corrective actions may be impeded or rendered ineffective if based on improper data. Chapter 6 contains additional specific information on labeling of samples and coordination with the laboratory.

The discovery of a sampling collection or handling problem is embarrassing if the errors are results of poor planning or poor technique. The following section discusses some common field sampling problems and offers suggestions on how to eliminate them.

Sampling equipment

As noted earlier, certain automatic samplers can introduce contaminants to the sample. These contaminants can result from the atmosphere, lubricants, or the materials from which the sampler is constructed. Sampler units should be checked by cycling distilled/deionized water through the sampler and analyzing the water for various constituents. If ambient air is used for purging cycles, the air source should be filtered to remove contaminants. All parts of the sampler, including purging chambers, pumps, and hoses, should be cleaned regularly. The frequency of cleaning depends on what is being sampled and how rapidly the unit gets dirty. Hoses should be clean and free of as much inert material as possible. Follow the manufacturer's recommended operation and maintenance procedures. Additional details on sampling equipment are given in Chapter 3.

When a sampler is relocated from one sampling point to another, it should be, at a minimum, cleaned thoroughly. Ideally, all hoses or parts coming in contact with the sample should be replaced. Setting up dual samplers in the same location can help identify a problem with sampling equipment.

Manual sampling equipment should be kept clean using detergents, acid soaking, and thorough rinsing. The equipment should be rinsed in the material being sampled before the sample is collected and rinsed in clean tap water after the sampling is complete. Ideally, specific sampling equipment should be designated and labeled for each sample location.

The choice of sample container type, size, and material depends on several considerations, including the volume of sample required, interference problems anticipated, type of testing to be performed, cost and availability, and resistance to breakage. The container requirements should be specified as part of the sample program. To prevent contamination, sample containers must be cleaned before being filled. New containers are not necessarily clean and must also be washed using prescribed procedures. At a minimum, all containers and caps must be washed in a non-ionic and nonphosphate detergent, rinsed well with hot tap water, and then rinsed with distilled water. Preferably, after being rinsed in distilled water, the sample containers and caps should be soaked in an acid solution at a temperature of approximately 70°C for approximately 24 hours. Quartz, polytetrafluoroethylene, or glass containers can be soaked in 1+1 nitric acid, 1+1 hydrochloric acid, or aqua regia (four parts hydrochloric acid to one part nitric acid) solutions. Plastic containers should be soaked in 1+1 nitric acid or 1+1 hydrochloric acid solutions. Solvent rinses are required for grease and oil or pesticide sample containers. These should be rinsed with hexane, followed by acetone, and finally with distilled water. Some container manufacturers certify that their containers are "clean," thus to be used without washing or sterilizing once a new case is purchased.

Because oil and grease in wastewater easily adhere to a number of materials, the recommended sample container is a wide-mouth jar with a polytetrafluoroethylene-lined screw. The wide mouth allows the technician to wipe the inside walls of the jar with the solvent used in the analysis. The jar should be precalibrated because the initial steps of the procedure are performed in the container.

Container caps or lids, which can potentially contaminate the sample, can be lined with aluminum foil or polytetrafluoroethylene before being placed on the container.

Sample containers should be prelabeled with the appropriate information (detailed in Chapter 6) and any special information the technician may need regarding the sample, such as unusual odors, appearance, or conditions that may have affected the sample. Sample containers that contain a preservative are not typically used to collect the sample. When a grab sample is obtained, the preservative may be added to the container before the sample is taken. A separate collection device should be used at each sampling location. With the exception of containers to which a preservative has been added and those used for oil and grease, containers should be rinsed with the sample material before collecting the sample. Oil and grease sample containers cannot be rinsed with the sample because of the container preparation and the adherence of oil and grease to the container walls. Sample containers or collection devices that are permanently stained should be discarded.

REAGENTS

Chemical contamination from reagents is a serious problem that results from either contaminated reagents or from adding the wrong reagent to a sample. Whenever a chemical's integrity is in doubt, it should be disposed of safely and properly.

The addition of incorrect chemicals to a sample and the cross-contamination of samples can be avoided by thorough planning and careful sample handling. The sampling plan designates the appropriate chemical preservative for specific sample containers. As noted earlier, containers containing preservatives should be clearly identified.

Chemical-dispensing equipment, usually a manual or automatic pipette, must be identified and labeled so that each is used only for one chemical. Using one pipette for zinc acetate and then nitric acid, for example, can lead to cross-contamination. The cleaning of dispensing equipment generally is the same as that for sample containers. The potential for ion exchange and leaching from container walls and plastic materials, discussed earlier, also applies to dispensing equipment and should be considered in the sampling plan.

Corrections can be made for sample contamination resulting from materials not removed by cleaning or the addition of reagents. Providing the laboratory with a "blank" container along with the sample allows the contamination levels to be determined. A "blank" container is one that has been treated in exactly the same manner as all other sample containers (for example, cleaned, taken to the field, and treated with preservatives). In the laboratory, distilled water added to the container is then used as a sample and passed through all of the analytical processes. The results can then be used to determine the level of external contamination. The use of blanks is described in Chapter 4.

*T*RANSPORTING SAMPLES

Once samples have been collected, they must be delivered to the laboratory. Whether delivered to an in-plant laboratory or transported a long distance, a minimum amount of information must accompany each sample bottle. This labeling information (Chapter 6) includes the sample source and location, date and time collected, preservatives added, analyses to be performed (if known), name of collector, type of sample, and remarks.

If possible, labeling information is attached to the sample bottles using an appropriate tag or label. Most facilities performing routine sample collection on an ongoing basis provide adhesive labels or tags that are attached to the sample bottles after the information has been filled in (Chapter 6). In addition, any specific or suspected problems encountered during sample collection that could affect the analytical results should be passed along to the laboratory.

Internal treatment plant sampling, in which the in-plant laboratory receives the sample, lends itself to scheduled collection and delivery. However, laboratory personnel must be aware of when samples are delivered. Samples not physically received by laboratory personnel stand a chance of being overlooked, lost, or delayed in analysis. The proper authorities should always be notified when samples have been delivered. Custody records that require signing when samples change hands are beneficial (see Chapter 6). The custody record should remain with the sample during analysis and be filed with the permanent copy of the laboratory results. In some cases, photographic evidence is needed at the time of sampling. If this is the case, proper identification of the photo and photographer is necessary. Information on the back of the photo should include a description of the subject, the date and time, and the signature of the photographer.

Transporting samples long distance to an outside laboratory takes some planning and prearrangement. Before samples are shipped, the following information must be provided to the laboratory supervisor:

- When the samples will be shipped;
- By what mode of transportation they will arrive;
- When they are scheduled to arrive;
- How many samples are being included; and
- What analyses are to be performed.

To satisfy these requirements, it will be necessary to make prearrangements with the shipper. Time schedules, fee payment arrangements, and container restrictions must be determined before plans can be completed. It is the responsibility of the person shipping the samples to ensure that they arrive in a timely manner and in good condition. To avoid breakage or leakage, it is necessary to provide sample containers that have enough head space to allow for temperature effects on sample volume. All caps must be securely in place and checked for leakage before packing; sealing caps with electrical tape is a common method for preventing leakage. Proper packing materials, such as styrofoam linings, must be used to protect glass containers. As an added protection to prevent leakage during shipment, samples and ice may be placed in a secondary plastic bag. The Department of Transportation restricts the transport of hazardous materials, including dry ice in air cargo. However, the hazardous materials regulations do not apply to certain materials (see Table 5.1, footnote c).

PRESERVATION TECHNIQUES

Guidelines have been established for the preservation of water and wastewater samples. Techniques discussed in this chapter are intended to provide an overview of options available for the preservation of samples. Additional information can be obtained from specific procedures described in *Standard Methods for the Examination of Water and Wastewater* (APHA, 1995). The legal requirements for sample preservation are contained in 40 CFR 136.

Refrigeration, pH adjustments, and chemical additions are the primary methods recommended for sample preservation. The methods listed in

---

Table 5.1 are presented as guidelines and may differ somewhat from those in other publications. Again, the sources listed above should be referred to for more specific information and legal requirements.

The most frequently used means of sample preservation is refrigeration at temperatures near freezing (4°C) or the use of wet ice. Biological activity is decreased because respiration rates are greatly reduced at low temperatures. Chemical reaction rates and the loss of dissolved gases also are reduced. It is often necessary to combine chemical addition or pH adjustment with refrigeration to ensure effective preservation.

Refrigeration at temperatures at or below freezing (0°C) is an effective long-term preservation method for only some parameters. In addition to killing or slowing the metabolism of microorganisms, the freezing of samples has other limitations. To avoid breakage, sample containers must have sufficient head space to allow for the expansion of freezing liquid. Particulate material will solubilize readily while samples are thawing because cell structures rupture when frozen. Samples to be analyzed for suspended solids should never be frozen.

Acid addition is a common method for decreasing both biological and chemical activity. Sulfuric acid (H_2SO_4) is added to a sample to stop bacterial action; preserving samples for chemical oxygen demand and organic carbon analyses are examples of when this application might be used. Sulfuric acid addition, combined with refrigeration, both preserves and pretreats oil and grease. Acids dissolve particulate matter and, therefore, must be avoided if suspended solids are to be determined. Alkali, typically sodium hydroxide (NaOH), is added to samples to prevent the loss of volatile compounds through the formation of a salt. Organic acids and cyanide are examples of salt-complexing compounds.

Chemicals, including the acids and bases discussed in the preceding section, are added to samples to stabilize compounds or stop biological activity. Copper sulfate ($CuSO_4$) and mercuric chloride ($HgCl_2$) are two biological inhibitory chemicals commonly used. Zinc acetate [$Zn(C_2H_3O_2)_2$], phosphoric acid (H_3PO_4), and sodium hydroxide (NaOH) are frequently used as complexing agents. The actual addition of chemicals alters the original composition of the sample. It is most important to avoid adding chemicals that contain elements for which the sample will be analyzed. For example, samples being collected for nitrogen analyses should not be preserved with nitric acid; samples for sulfate should not contain sulfuric acid. In many instances, the volume of preservative must be taken into account to properly establish the concentration of contaminants. The combinations of possible cross-contaminations are great; however, they can be avoided by using common sense and carefully planning the sample program.

Sample preservation can be replaced with an alternative procedure with the proper regulatory approval as outlined in 40 CFR 136.

Biomonitoring is currently an important tool for monitoring facility operational performance. Preservation for this testing is not covered in 40 CFR 136. Containers should be plastic or glass. Samples should be cooled to 4°C, and the holding time is 36 hours.

REFERENCES

American Public Health Association (1995) *Standard Methods for the Examination of Water and Wastewater.* 19th Ed., Washington, D.C.
U.S. Environmental Protection Agency (1993) Title 40, 40 CFR 136. U.S. Code of Federal Regulations.

SUGGESTED READINGS

U.S. Environmental Protection Agency (1982) *Handbook for Sampling* and *Sample Preservation of Water and Wastewater.* EPA-600/4-82-029, Washington, D.C.
U.S. Environmental Protection Agency (1989) *POTW Sludge Sampling and Analysis Guidance Document.*
U.S. Environmental Protection Agency (1990) Sampler's Guide to the Contract Laboratory Program.
Water Environment Federation (1990) *Operation of Municipal Wastewater Treatment Plants.* Manual of Practice No. 11, Alexandria, Va.

Chapter 6
Data Collection and Recording

INTRODUCTION

It is the responsibility of all personnel collecting or reporting such data to recognize the critical nature of proper data management. Personnel collecting the sample and field data and reporting the findings should be encouraged to take personal ownership of their responsibilities.

This chapter will serve as an overview of data collection and recording from setting up a data collection program to possible subjection to a regulatory audit. It will cover proper sample labeling as well as instructions for laboratory and sampling personnel. It should be understood that although every program can be different and requirements will differ, the fundamentals change little. These basic guidelines are followed from program to program and will make a notable difference in how well a data collection program is understood by others.

SETTING UP A PROGRAM

When setting up a data collection program, it is first important to designate a program coordinator. This person must understand what samples or analytical data are requested or required. A program coordinator should also be aware of whether the data will be used for process control, regulatory requirements, or other reasons. Additional factors such as budgetary allotments, equipment needs, and weather patterns will also have to be addressed. Once the coordinator realizes the scope of the program, a team can be formed.

The team should have representatives from each discipline as defined by the scope of the program. Representatives of the analytical laboratory, quality control, the sampling team, the treatment plant, and safety and regulatory groups must be considered. This team should then forward any pertinent information or instructions to the personnel performing the required tasks.

All members of this team will need to decide if further training or procedure updates are necessary and whether any reviews or extra training must be conducted in their respective areas. In some cases, formal documentation of such training may be required.

Because of the high costs of data collection and recording and analytical testing, all participants must clearly understand their roles in safe and accurate sample and data collection and recording.

In some programs, a sample collection schedule will have to be generated between the laboratory and sampling personnel. Should it become necessary to ship samples to an "outside" or contract laboratory, a sampling schedule will be critical to transportation and work load planning. Smaller, "in-house" programs tend to be more flexible, but a sample collection schedule should still be considered.

INSTRUCTIONS TO SAMPLING PERSONNEL

The individuals whose task it is to collect samples and record field data will play major parts in contributing "hands-on" input to the program. Some smaller programs may have one or two people collecting samples and recording and reporting the data. Communication in these smaller programs will be less difficult than in large programs, where groups of people are performing different tasks. In either case, it may be necessary for the coordinator to meet with the sampling personnel and explain the goals of the project. At this point, the program coordinator and sampling personnel should be ready to discuss

- Safety:
  - — Personnel protective equipment—follow Occupational Safety and Health Administration recommendations.
  - — Clothing—address and correct any concerns.
  - — Specific site hazards—tour area to be sampled.
- Regulatory requirements (if applicable):
  - — State and federal compliance.
  - — Local regulations.
- Equipment needs:
  - — On-site electrical.
  - — Shelters.
  - — Materials of construction.
- Sample handling:
  - — Chain of custody required.
  - — Labeling.
  - — Scheduling.
  - — Preservation.
  - — Sampling containers.
- Field data collection and data recording:
  - — Log books.
  - — Maps.
  - — Flow charts.
- Transportation and delivery of samples:
  - — Destination of samples and/or data.
  - — Receiving personnel.
  - — Log in personnel.

A representative from the laboratory should help answer some of the questions or concerns that may arise concerning analytical needs. In some cases, it will be necessary to perform certain testing at the point of sample collection. A lab representative may often be the best source for training sampling personnel who might be performing on-site analyses.

SAMPLE LABELING

The sample label is one of the primary forms used for recording field data. The choice of label could depend on two determining factors: if the sample container will be cleaned, sterilized, and reused, then a label that is removable should be used; if the sample containers will be discarded, then a label with stronger adhesive should be affixed to the container.

The following information should be included on the sample label:

- Location identifier—predetermined location I.D. code or name;

- Date(s) of sample collection;
- Time of collection—on and off times;
- Name or initials of sampling personnel;
- Type of sample collection—composite or grab, for example; and
- Preservative(s) added.

Labels that can be handwritten using indelible ink can be made with standard masking tape, or blank removable labels may be purchased. Blank labels often are available in a variety of sizes. Rolls of blank labels may also be used in some office printers, allowing the facility to design its own label. These labels are typically removable and can be used on most container types. Preprinted labels from a supplier generally have a stronger adhesive but tend to be more expensive. Preprinted labels allow the person collecting samples to fill in the blanks with the data needed to calculate sample results.

When applicable, information can be transcribed directly to the container, such as when using a permanent marker on polyethylene jugs.

Prespecified bar code labels are manufactured by vendors for use on specific sample containers. This labeling method leaves a tracking trail from the container's manufacture to the disposal of the container after its use. Computer instrumentation is required to implement this labeling system, so budgetary concerns will have to be considered.

Regardless of the type of label used, it should be easy to work with and fulfill the following guidelines:

- Adheres to the container properly—the label cannot fall off when sample is collected.
- Removable—if sample container is to be reused.
- User friendly—allows enough space for required data and information to be legibly recorded; works well in inclement weather.

In some sampling situations, it may be wise to collect the sample first and then attach the sample label (do not place labels on cover or caps). This will allow the person collecting the sample to clean off any sample residue that may have collected on the sides of the container.

To record the field data that is required on the sample label, use a permanent ink pen. Most pens will work on labels that are clean and dry. A permanent ink, felt-tip, fine-point marker typically works well, except when the label is wet because of weather conditions or condensation from the sample.

The ink should be able to dry without smudging. Also, the ink should not be allowed to leach through the sample container. Writing directly on plastic sample containers, for instance, could result in contamination of the sample.

The data recorded on the label should be neat and legible. Any mistakes made on the label should be crossed out once and initialed by the person doing the sampling. This saves time when transferring data from the label to the selected data system. When samples are shipped and analyzed at a contracted laboratory, an illegible label could hold up the analytical process or be questioned during an audit. Trying to figure out the information and data on a poorly prepared sample label can be time consuming and costly.

Additional information such as temperature, pH, or total chlorine count may also be noted on the label. Regardless of the information found on a label, any mistakes must be crossed out with one line and initialed by the person making the correction.

After the sample is in the container and the label is properly affixed, the sample can be transported to the laboratory. Some samples must be transported in a refrigerator or a cooler packed with ice. In situations for which ice is used, it may be helpful to put a sheet of plastic between the ice and the sample container or place sample bottles in individual plastic bags. This will protect the label from getting wet and will allow further data to be added at the laboratory if needed.

*L*ABORATORY IDENTIFICATION

After delivery to the laboratory, the sample or data readings will have to be entered to the selected tracking system. The laboratory personnel may simply request a hard copy chain-of-custody form to receive the samples for analysis.

The chain-of-custody form (Figure 6.1) is an important document and must be kept totally legible and on file for the amount of time determined by internal policy or permit requirements. Often, the chain-of-custody form accompanies the sampling personnel during the sampling activity and, therefore, is subject to many different climates. Because this document must be kept legible, all necessary precautions should be taken when handling it. A black hard-point ink pen is preferred for filling out the chain-of-custody form because it allows for better microfiche records and copies. During an audit, it is common for the auditor to ask to inspect chain-of-custody documentation for clarity and completeness. If filled out correctly, this documentation should track a sample from collection to storage or destruction of the sample.

Examples of electronic chain-of-custody forms, both blank and completed, are shown in Figure 6.2. These documents will follow the sample from person to person and analysis to analysis. Other data-tracking systems might require the sampling or laboratory personnel to "enter" or "log in" the sample taken and related data to a computer database. The computer should assign a tracking number to the sample or data reading after the associated log-in codes are supplied via the computer's log-in screen. Examples of blank and completed sample log-in screens are shown in Figure 6.3. These codes will notify the laboratory of the source of the sample and data; the personnel who collected the sample; analyses to be performed; and other pertinent information. As the analytical results are "entered" or "posted" in the computer, the

---

SURVEY NO. SAMPLE NO.		DATE COLLECTED		Day/Month/Year	COLLECTED BY		(Name)
NUMBER OF SAMPLES		TYPE OF SAMPLE	Raw Water Influent Bay Water Effluent		Industrial Waste Other:		

Sequence No.	Station	Type Sample		Analyses Required	Relinquished by	Date/Time
		Comp.	Grab			
					Received by	Date/Time
					Relinquished by	Date/Time
					Received by	Date/Time
					Relinquished by	Date/Time
					Received for Lab by	Date/Time
Remarks (See Reverse)					Mailed BY DATE	

SURVEY NO. SAMPLE NO.	NUMBER OF SAMPLES	DELIVERED TO	DELIVERED BY DATE

Figure 6.1   Typical chain-of-custody form.

sample tracking is electronically recorded. The computer will show where the sample was, who performed the analyses, and the sample turnaround time. Sample results are not the only data that may be recorded in the computer. Different field test results and information such as temperature, pH, and total chlorine count can also be recorded. These data can be assigned their own tracking numbers, or existing numbers from a sample that has already been logged in could be used for querying or reporting purposes.

SAMPLE AND DATA SPREADSHEETS

Another option for sample and data recording is the sample and data spreadsheet (see Figure 6.4). The spreadsheet is typically an 8.5 in. × 11 in. or 11 in. × 17 in. sheet of paper broken into columns of required data and analyses. One column will provide a predetermined sample or data-tracking number. Another column is provided for the name or initials of the personnel reporting the required data and information. The spreadsheet may be kept in a labeled folder or hard-cover office binder. A designated person assigns the sample or data reading its own "log-in" number. As the analyses are per-

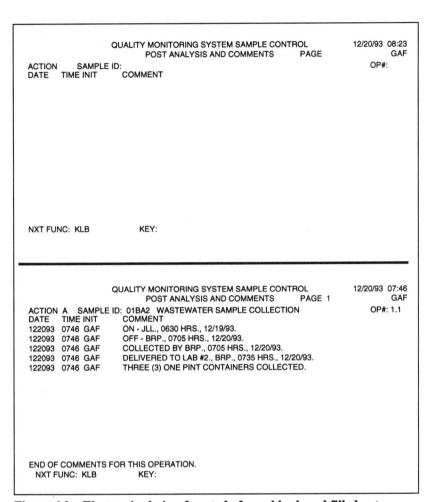

```
                    QUALITY MONITORING SYSTEM SAMPLE CONTROL          12/20/93  08:23
                         POST ANALYSIS AND COMMENTS       PAGE            GAF
   ACTION      SAMPLE ID:                                              OP#:
   DATE    TIME INIT     COMMENT




   NXT FUNC:  KLB           KEY:
```

```
                    QUALITY MONITORING SYSTEM SAMPLE CONTROL          12/20/93  07:46
                         POST ANALYSIS AND COMMENTS       PAGE  1          GAF
   ACTION  A   SAMPLE ID: 01BA2  WASTEWATER SAMPLE COLLECTION           OP#: 1.1
   DATE    TIME INIT     COMMENT
   122093  0746  GAF     ON - JLL., 0630 HRS., 12/19/93.
   122093  0746  GAF     OFF - BRP., 0705 HRS., 12/20/93.
   122093  0746  GAF     COLLECTED BY BRP., 0705 HRS., 12/20/93.
   122093  0746  GAF     DELIVERED TO LAB #2., BRP., 0735 HRS., 12/20/93.
   122093  0746  GAF     THREE (3) ONE PINT CONTAINERS COLLECTED.




   END OF COMMENTS FOR THIS OPERATION.
      NXT FUNC:  KLB           KEY:
```

Figure 6.2   Electronic chain-of-custody form, blank and filled out.

formed, the results are recorded in pen on the spreadsheet corresponding to the assigned log-in number. Again, any mistakes made on a spreadsheet should be crossed out once and initialed by the person responsible for the mistake.

*F*IELD DATA RECORD BOOKS

In some sampling programs, it may be suggested or required to document the field data and observations in field data record books, which are commonly available. These books are typically hard cover, with numbered pages. A permanent ink pen should be used for recordings and documentation. Any mistakes made in the field record book should be crossed out with a single line and initialed by the person making the record entry. No pages should ever be

```
                    QUALITY MONITORING SYSTEM SAMPLE CONTROL
                              ENVIRONMENTAL LOGIN
LOCATION:
TYPE:            1=ENVR      2=ENVS
SUBTYPE #:       1=WATER  2=AIR  3=SOLIDS  4=GRAB  5=24HR  6=48HR  7=7DAY  8=8DAY  9=UTIL
SPECIFIC SUBTYPE:                          * * * SUBTYPE VALUE IF LEAVE SUBTYPE # BLANK * * *
ROUTE ID:
DATE SAMPLED:           /      /      TIME SAMPLED:      SAMPLER:      SAMPLE TYPE:
NEW SAMPLE ID:
- - - - - - - - - - - - - - - - - - - - - - - - - - - - - - - - - - - - - - - -
OPTIONALLY ENTER THE FOLLOWING INFORMATION TO OVERRIDE THE DEFAULT VALUES.



            REMARKS:


TITLE:

REPORT TO:                          EXT #:          CHARGE #:

NXT FUNC: KLB           KEY:




                    QUALITY MONITORING SYSTEM SAMPLE CONTROL
                              ENVIRONMENTAL LOGIN
LOCATION: ZZ-1
TYPE:       1    1=ENVR     2=ENVS
SUBTYPE #:  5  1=WATER  2=AIR  3=SOLIDS  4=GRAB  5=24HR  6=48HR  7=7DAY  8=8DAY  9=UTIL
SPECIFIC SUBTYPE:                          * * * SUBTYPE VALUE IF LEAVE SUBTYPE # BLANK * * *
ROUTE ID: 10
DATE SAMPLED:        12  /  19  / 93     TIME SAMPLED: 0705  SAMPLER: BRP   SAMPLE TYPE:
NEW SAMPLE ID:
- - - - - - - - - - - - - - - - - - - - - - - - - - - - - - - - - - - - - - - -
OPTIONALLY ENTER THE FOLLOWING INFORMATION TO OVERRIDE THE DEFAULT VALUES.
DOCUMENT ID :              JOBF:


            REMARKS: Sample contained many solids.


TITLE:  Wastewater Sample Collection

REPORT TO:  Smith                   EXT #:          CHARGE #: 3321

NXT FUNC: KLB           KEY:
```

Figure 6.3   Sample log-in screens.

removed from the book. If, for any reason, a page cannot be used, the words "Not Used" should be printed clearly across that page and initialed. Each documented entry in these books should include the monitoring location, date, time, names of personnel, and other important information such as routine calibrations and needed maintenance.

CHART RECORDING

Many types of instrumentation use charts to record or track information. Although there are too many types of instrument charts to cover in this chapter, the information is relatively similar for each. Upon visually inspecting the instrument and chart in question, the person responsible should use an ink pen

Daily Sample Log-In List

Month _____

Sample Identification Number	Date	Time	Location	Type	pH	Temp (C)	Flow	Initials
A-0100								
A-0101								
A-0102								
A-0103								
A-0104								
A-0105								
A-0106								
A-0107								
A-0108								

Figure 6.4   Data spreadsheet.

to record his or her name or initials, the totalizer reading (if necessary), time and date, and name of location. Of course, other information could also be recorded on these charts depending on what is required.

*I*NTEGRATED SAMPLE ANALYZERS AND DATA

A relatively new technology integrates instruments that analyze the samples and store them in a database. Examples include autosamplers coupled to heavy metal analyzers, gas chromatographs, or mass spectrometers and may include on-line analyzers as well. There are advantages and disadvantages to integrated systems that should be considered. A major determining factor in selecting this type of system might be budgetary concerns. This instrumentation tends to be expensive and requires costly maintenance. Laboratory space limitations will also help determine the practicality of such instrumentation. Another factor is the tendency for an instrument to foul when used to analyze a variety of wastewater samples. Proper cleaning, calibration, and standardizing are critical when using these instruments. A guideline of specifications on a given constant waste stream allows the operator to watch for drifts in calibration or other related concerns that may arise. When an instrument is used to analyze a variety of waste streams, the operator must be more sensitive to possible drifts in the accuracy of the instrument and the data being recorded.

The advantages of properly calibrated and maintained automated data collection systems are many. When budgetary and laboratory evaluations allow the purchase of such instrumentation, it can be a valuable tool. One advantage may be a reduction in the number of employee hours spent manually handling each sample to be analyzed. With some integrated systems, a number of samples are loaded into a sample rack and the instrument is programmed to perform the analysis. Quicker turnaround time on sample analyses means less lag time when making important changes to processes. This, in turn, could reduce operating costs or prevent an environmental upset.

RECORDKEEPING

In all aspects of wastewater sampling, data collection, sample analyses, and instrument calibration, recordkeeping is useful and sometimes required. Records may consist of calibration charts, calibration procedures, factory specification manuals, and strip charts from instruments used in wastewater analyzers. The period of time these records must be retained varies depending on the scope of the program. Where data collection is mandated by environmental regulations, record retention can vary from 1 year to the life of the program. In smaller, in-house programs, record retention will depend on the needs of the program leader and the future application of the information that has been collected. A good recordkeeping system is, in many ways, as important as collection and analysis of the samples themselves. Records should always be kept secure, neat, and orderly. Sensitive or confidential data may need to be kept under lock and key to provide proper security. A sign-out log for the custody of these documents and data should also be considered. Records can also be microfiched to aid in storage space.

AUDITS

The word "audit" is often misconstrued to mean that a program is deemed to be in disarray. This is not the case unless the program has failed to document properly and follow correct procedures. An audit can be a good measuring device to find out if some points of a program need to be upgraded. Many times, an audit can help with a new idea on how to improve sample collection and delivery, make laboratory personnel aware of new analytical techniques, or provide advice on proper record and data management. An audit can be performed by a team of people from within the program or may be requested by a regulatory agency.

Regardless of who conducts the audit, it is important to always be prepared. Although 24 hours is usually the minimum amount of notice given by an audit team, unannounced inspections can be made. Before any audit or inspection, a policy should be in place to cover such requests. This policy should

- Identify activities that are subject to federal, state, and local regulations;
- Identify which cases require legal counsel;
- Define appropriate employee behavior (cooperative, courteous, and candid);
- Address all sensitive and confidential data and security of these data; and
- Be familiar to the staff, with on-going revisions as required.

Upon notification of an audit, it is necessary to identify

- Escorts and alternates who are knowledgeable of the work place, pleasant, and have effective communication skills.
- Ground rules of the inspection. These should include a "statement of intent" outlining what is to be inspected, the purpose of the inspection, and other rules or regulations pertaining to the company's policy for audits. Ground rules should also address any legal concerns that may arise during the inspection.

A few additional points of interest to the audit team could be the following:

- Are all personnel training records up to date?
- Are all analytical testing instruments calibrated per factory specifications and available calibration records?
- Are sample labels legible, and is all necessary information present?
- Are samples being collected, handled, and transported according to approved procedures (see U.S. EPA, 1992, and ISO, 1990).

- Are all sampling, laboratory, and recordkeeping procedures up to date and available for review?

This list is not comprehensive but gives a basic idea of what may be of interest to an audit team. To avoid a long, drawn out, and possibly poor review, it is important to have everyone trained properly and fully understanding what is required of them and why. Hold internal audits periodically to ensure that the basic rules, regulations, and procedures are being followed. Questions that could arise throughout the program may need to be researched and educated answers provided in a timely manner. If all personnel are properly trained and understand the requirements of the program and their respective responsibilities, an audit can be a good opportunity to show others what a well-run program can achieve.

*R*EFERENCES

International Organization for Standardization (1990) *Standard Methods for General Requirements for the Competence of Calibration and Testing Laboratories.* 3rd Ed., Geneva, Switz.
U.S. Environmental Protection Agency (1992) *Good Laboratory Practice Standards Final Rule.* Office of Solid Waste, Washington, D.C.

Chapter 7
Data Preparation

*I*NTRODUCTION

This chapter provides information on preparing data gathered from wastewater sampling. It discusses how to perform simple statistical calculations to determine the quality of data and how to perform calculations using these data to arrive at conclusions. Regulatory requirements are also discussed.

As the preceding chapters have indicated, a great deal of time and care must be expended in collecting, preserving, and transporting samples to ensure that they are representative of the waste stream being sampled. The chief purpose of this effort is to provide the most accurate data possible for making decisions. Depending on the purpose behind the sample collection, the data may be intended for process monitoring, process control, evaluating process performance, analyzing future design requirements, or even instituting enforcement action. Under normal circumstances, the generated data will be received either in the form of data points or as a graphical output from an on-line, continuous-monitoring system. The purpose of this chapter is to present information and techniques that will reduce these raw data to a more manageable and useful form, while at the same time ensuring that the information is neither masked nor altered.

Techniques for evaluating the accuracy and dependability of the analytical data are also discussed in this chapter, and the relative merits of several methods for determining whether to discard a piece of data will be compared. The material presented early in the chapter applies to anyone working with data collected for laboratory use. More complex calculations and guidance for preparing material for scientific publications follow the early discussion.

*R*EGULATORY REQUIREMENTS

This section characterizes the pertinent federal requirements and guidelines that apply to data preparation for subsequent recordkeeping and analysis. Much of the data generated and analyzed at wastewater treatment plants is recorded on forms and submitted to the U.S. Environmental Protection Agency (U.S. EPA) or state agency as part of a discharge permit such as a National Pollutant Discharge Elimination System (NPDES) permit. The data can also be used to generate reports that are, in turn, submitted to the same agencies. This section will address the applicable guidelines of these agencies, including:

- National Pollutant Discharge Elimination System forms, and
- Compliance inspection requirements.

DISCHARGE MONITORING REPORTS. Certain regulatory standards are established by agencies such as U.S. EPA. The seriousness with which U.S. EPA considers the data reported from wastewater treatment plants is reflected in the certification statement found on a monthly discharge monitoring report (DMR), U.S. EPA Form 3320-1 (significant portion in italics):

> I certify under penalty of law that I have personally examined and am familiar with the information submitted herein, and based on my inquiry of those individuals immediately responsible for obtaining the information, *I believe the submitted information is true, accurate, and complete.* I am aware there are significant penalties for submitting false information including the possibility of fine and imprisonment.

Discharge monitoring reports are report forms prepared by U.S. EPA and used by permittees to report various parameters required by their NPDES discharge permits, such as effluent biochemical oxygen demand (BOD), plant flow rate, and total suspended solids. These parameters may be limited by the permit on a daily, weekly, or monthly basis, so the person responsible at the facility must have an understanding of the mathematical analysis necessary to complete the DMR properly. By signing the DMR, the responsible person, whether the chief operator or other person in the organization, is certifying that the data shown are correct.

U.S. EPA has developed guidelines to be followed to complete a monthly DMR properly. It is important for the responsible person to fully understand these guidelines. The responsible person must follow the guidelines from his or her own regulatory agency, which may differ from information explained in this chapter. For instance, the 7-day average for effluent BOD is defined by U.S. EPA as

> ...the highest average calculated for any consecutive 7-day period within the calendar month.

As defined later in this chapter, this is called a rolling 7-day average within a calendar month. Some agencies may define permit monitoring requirements for a weekly average, which is the average of data within a 7-day calendar week, from Sunday through Saturday.

COMPLIANCE INSPECTIONS. In addition to requiring a DMR, U.S. EPA, or a designated state agency, conducts periodic facility inspections. The various types of inspections are defined in the *NPDES Compliance Inspection Manual* (U.S. EPA, 1988). They include compliance evaluation, compliance sampling, toxics sampling, compliance biomonitoring, performance audit, diagnostic, pretreatment compliance, reconnaissance, and legal support inspections. Of particular importance to data preparation, the following items are found in U.S. EPA's inspection checklist for data handling and reporting:

- Round-off rules uniformly applied,
- Significant figures established for each analysis,
- Provision for cross-checking calculations used,
- Correct formulas used to calculate final results,
- Control chart approach taken in calculating final results,
- Report forms developed to provide complete data documentation and permanent record and facilitate data processing, and
- Data reported in proper form and units.

Each of these items and numerous others are reviewed by the agency inspector and checked off as either "yes" or "no." If a "no" response is given, then corrective action will be taken or required. Thus, treatment plant personnel must be aware of the importance of correctly handling data from laboratory measurements and field instrumentation.

DATA REDUCTION

This section provides information on data reduction and encompasses significant figures and the rounding off of figures. Also included is elementary information that can be gleaned from a set of related data points, such as the median, range, and percentile. Table 7.1 lists typical monthly data collected at a wastewater treatment plant.

SIGNIFICANT FIGURES. The number of digits in which a piece of data is reported is an indication of the level of accuracy of the data. Reporting more digits than are significant, therefore, can lend a greater degree of accuracy to a data point than it warrants; whereas failure to carry all significant figures out will result in a loss of meaningful information.

Recording Numbers. Record all digits that are known to be correct plus the next one in which there is a degree of uncertainty. Thus, recording the number 23.6 indicates that the value is known accurately to the units place (the 3), but that there is some doubt about the first decimal place (the 6). In this case, the data point has three significant figures.

The accuracy of the data point is determined from either the laboratory accuracy for a laboratory test result (such as a BOD test result) or the accuracy of an instrument that directly measures a parameter (such as a wastewater flow meter). Accuracy is discussed later in this chapter.

For example, if the previous value of 23.6 is known to be ±1% accurate, then a reading could actually range from 23.4 (23.6 – 1% of 23.6) to 23.8

Table 7.1   Typical raw data.

Date	Flow rate, mgd[a]	Raw pH	Biochemical oxygen demand, mg/L	Total suspended solids, mg/L	Effluent pH	Carbonaceous biochemical oxygen demand, mg/L	Effluent biochemical oxygen demand, mg/L	Effluent total suspended solids, mg/L	Chlorine residual, mg/L	Effluent fecal coliform density, MPN[b]
1	23.6	7.2	272	182	6.9	8.0	12	18.4	0.6	2
2	24.0	6.6	379	226	6.7	6.6	11	21.1	0.9	
3	24.4	6.6		192	6.7			17.2	0.9	2
4	24.0	6.6	311	184	6.9	4.0	9	16.5	0.9	
5	23.8	6.7	291	175	6.8	6.8	9	16.0	0.8	
6	22.9	6.8	311	186	6.8	6.6	9	15.8	0.9	
7	22.0	6.8	226	168	7.0	6.2	10	16.8	0.9	
8	22.9	7.2	274	154	7.0	4.4	9	16.6	0.9	0
9	24.1	6.8	371	148	6.8	5.9	11	16.0	0.8	
10	23.5	6.7	358	184	6.8	5.5	9	13.6	0.7	0
11	25.0	6.8	302	176	6.9	4.0	15	15.4	0.9	
12	22.4	6.7	336	196	6.9	6.2	9	15.8	0.8	
13	19.9	6.7	296	168	7.3	5.1	10	15.0	0.8	
14	20.0	6.6	248	153	7.3	4.8	9	12.8	0.9	
15	21.2	7.0	286	192	7.3	3.4	11	13.2	0.8	0
16	21.4	6.7	326	193	7.3	4.6	8	15.8	0.8	
17	25.5	6.9	332	190	7.1	3.8	9	14.6	0.8	0
18	25.1	6.7	322	184	7.2	4.0	10	13.4	0.8	
19	21.1	6.6	306	168	7.1	5.0	8	13.6	0.8	
20	20.5	6.8	313	174	7.2	5.3	8	12.8	0.9	
21	18.3	7.0	245	186	7.3	5.0	7	14.2	0.9	
22	24.0	7.2	234	162	7.3	6.4	9	13.8	0.8	0
23	25.2	6.7	320	176	7.1	4.7	8	15.8	0.8	
24	25.5	6.9	290	175	7.3	6.3	9	14.2	0.8	2
25	24.2	7.2	258	166	7.3	4.6	8	14.2	0.9	
26	21.0	7.1	207	150	7.2	4.3	7	14.6	0.8	
27	21.5	7.2	199	142	7.2	4.1	8	14.0	0.8	
28	22.6	7.0	206	158	7.0	5.5	8	14.2	0.6	
29	24.2	7.2	225	160	7.2	5.6	8	13.6	0.8	0
30	25.1	7.0	263	182	7.2	5.0	8	13.2	1.0	5
Total	689									

[a] $mgd \times (3.785 \times 10^3) = m^3/d$.

[b]   MPN = most probable number.

(23.6 + 1% of 23.6). Thus, the units place (3) is known accurately, but there is some doubt about the first decimal place (.6).

There are some basic rules pertaining to the determination of how many significant figures a recorded data point contains. The chief area in which questions can develop concerns the impact of zeros because they may or may not represent significant figures, depending on the circumstances.

Significant Figure Rules. In general, the following rules apply to significant figures:

1. All digits that are nonzero integers are significant. Thus, the number 20 has one significant figure, the number 24 has two significant figures, 23.6 has three significant figures, and so on.

2. Zeros bound on both sides by nonzero integers are significant. Thus, the number 1 001 has four significant figures, and the number 5 700 607 has seven significant figures.

3. Zeros after a decimal point and preceded by a nonzero integer are significant. Thus, the number 7.600 000 has seven significant figures, and the number 7.600 has four significant figures.

4. Zeros after a decimal point but not preceded by a nonzero integer are not significant but are simply indicating the position of the decimal place. Thus, 0.006 7 has two significant figures, but 0.006 70 has three significant figures.

5. Zeros following a nonzero integer and preceding the decimal point may or may not be significant. Thus, the number 1 000 may have from one to four significant figures, depending on whether the zeros simply are marking the decimal place or actually are significant. This evaluation will have to be based on the source of the data, the experimental accuracy, and other factors. As a special case, the number 1 000.0 can be assumed to have five significant figures, because the inclusion of the zero following the decimal place falls under rule 3. In the same manner, the number 100.000 0 has seven significant figures.

ROUNDING OFF. Once the number of significant figures of a data point is known, it is possible to proceed with the reduction of the data to a more usable form. In almost all actual cases, data being manipulated mathematically will consist of a series of values having different numbers of significant figures. Consequently, it is necessary to round off, at some point, to the appropriate number of figures.

Rounding Off Rules. The rules for rounding off are as follows:

1. If the number following the digit to be rounded off is greater than 5, the digit should be raised to the next number. Thus, 46.760 1 rounded to three significant figures would be 46.8 because 601 is greater than 500.

2. If the number following the digit to be rounded off is less than 5, the digit should be left unchanged. Thus, 46.764 99 rounded to four significant figures would be 46.76 because 499 is less than 500.

3. When the number following the digit to be rounded off is equal to 5, two widely used procedures apply:

— The accepted procedure from the U.S. EPA's *Handbook for Analytical Quality Control in Water and Wastewater Laboratories* (1979) is to change the number to be rounded to the nearest even digit. Thus, 32.475 00 rounded to four significant figures would be 32.48. However, 32.485 00 rounded to four significant figures would also be 32.48.

— Most hand-held calculators and desk-top computers will round digits followed by a 5 to the next digit. Thus, 32.475 rounded to four significant figures would be 32.48, and 32.485 would be 32.49, not 32.48.

— Generally, the difference caused by these two methods is insignificant except when rounding off to two significant figures. Readers accustomed to using desk-top computers and hand-held calculators should follow the methods used by their computers or calculators to be consistent.

The rules for rounding off a number depend on the kind of mathematical operation being performed. The end results of all such rounding-off operations is to produce a final number with no more significant digits than that of the least accurate number.

Rounding Off Rule During Mathematical Procedures. For addition, multiplication, and division, rounding off should be done only after the operation is completed; however, for subtraction, each number is rounded off before the operation. When a series of operations is being performed, it is best to avoid rounding off until the series of calculations is completed. Most modern calculators and computers will delay rounding off automatically, even though the display is rounded off.

REDUCTION OF RECORDED DATA. Modern electronic instrumentation renders a continuous signal reading such as a flow rate. In computerized data systems, only periodic readings, such as once every 15 seconds or once every minute, are recorded. In addition, modern flow meters total the readings on a daily basis. The daily average can then be determined from this total flow number.

Some data are presented in the form of a continuous recording rather than as discrete data points. Among the types of information that can be presented in this format are flow, dissolved oxygen, conductivity, and pH. An example of such a recording is a strip chart, which shows flow over a 24-hour period. The specific data available from such a recording include maximum and minimum values, instantaneous values at specific times, and daily averages.

Following is the procedure for obtaining a daily average from a recording:

1. Divide the period in question into a number of equal intervals. The accuracy of the average will increase as the number of intervals increases. A highly variable parameter will benefit from the use of a large number of intervals, whereas one that changes slowly can be evaluated accurately using fewer intervals. For flow, hourly intervals typically will be adequate.
2. Evaluate the average value for each time interval to the highest accuracy possible and record each value.
3. Sum all values generated in step 2.
4. Divide the sum from step 3 by the total number of intervals used.
5. Round the results from step 4 to the number of figures of the individual data points enumerated in step 2.

Numerous operations can be performed on a collection of data points to either modify or simplify them for evaluation. The operation performed will depend on the information desired. The common methods for modifying data include median, range, and percentile.

MEDIAN. The median (middle) value of a series of data points is the midpoint value of the series after all of the data points have been sorted in either descending or ascending order.

- For a series having an odd number of data points, the median is the value that has an equal number of data points both above and below it.
- For a series having an even number of data points, the median is the value that lies halfway between the two numbers that are bound by the same number of data points.

A median is sometimes useful for evaluating and defining data that are highly variable. The procedure for determining the median of a set of data is as follows:

1. List the data in order of increasing or decreasing value.
2. Determine the total number of data points. If there is an odd number of data points, subtract 1 from the total; if there is an even number of data points, subtract 2; in either case, divide the result by 2.
3. Count from either end of the ordered series until the number calculated in step 2 is reached. For a series having an odd number of data points, the next data point in the series is the median. For a series having an even number of data points, the sum of the next two data points divided by 2 is the median.

PERCENTILE. The percentile value is similar to the median except that, instead of being the middle value of a sorted series, the value is a certain percentage of the range of values. A percentile is a value on a scale of 100 that indicates the percent of a distribution that is equal to or below it. For example, a score of 95 is a score equal to or better than 95% of the scores. The following example uses Table 7.2 to determine the 90th percentile.

The median value from this data set is 13.5, which is the average of 13.1 and 13.9. The 90th percentile value is 15.6, which with 10 data points is the 9th value. Interpolation may be necessary in some cases to determine a percentile value. For example, assume that two additional data points are added to the previous set:

11	16.9
12	17.2

The 90th percentile value is for the "10.8th" value in the set (or 0.90×12). The "10.8th" value is found by interpolating between the 10th value of 15.8 and the 11th value of 16.9. Because 10.8 is 80% of the span between 10 and 11, the 10.8 value is 80% of the difference between 15.8 and 16.9, added to 15.8. Eighty percent of 1.1 is 0.88, which added to 15.8 equals 16.68. This should be rounded off to 16.7.

RANGE. A set of data can be characterized by its range (its maximum and minimum values). Range is frequently useful for characterizing parameters that can affect a treatment system adversely when they fall outside of a particular span of values: the range of pH is an obvious example. Under normal circumstances, range can be determined by a simple examination of the data.

Most spreadsheet software programs for desk-top computers have built-in functions to determine minimum and maximum values of a set of data. An example would be the range of flow rates shown in Table 7.1. The minimum

Table 7.2   Percentile example.

Sorted data number	Result
1	10.2
2	11.7
3	11.9
4	12.7
5	13.1
6	13.9
7	14.6
8	15.1
9	15.6
10	15.8

value of 18.3 is seen on the 21st of the month. The maximum value of 25.5 is seen on the 17th and the 24th.

ELEMENTARY STATISTICS

AVERAGES. The average, or mean, as the name implies, represents the average set of conditions over the period under consideration. In most cases, the combination of range and average is necessary to define the data adequately. Thus, knowing a treatment plant has discharged an average BOD concentration of 9 mg/L for a month is of little value in determining how the plant has performed on a day-to-day basis throughout the month. On the other hand, knowing the plant effluent BOD ranged from 7 to 35 mg/L provides better information concerning plant performance during the month but has little value for determining whether the plant met permit conditions.

With NPDES permits, the permittee is typically required to calculate daily, weekly, and monthly averages. The types of averages are generally arithmetic, but for some parameters, such as fecal coliform, a geometric average is required. In all cases, the purpose of an average is to provide a data point that is representative of the period under consideration. The most common averages, or means, are arithmetic, weighted, and geometric.

ARITHMETIC MEAN (AVERAGE). The most common average is the arithmetic mean. It is used when all data points are equally significant, such as when average monthly values are calculated. An arithmetic mean is calculated by taking the sum of all data points and dividing this sum by the total number of data points. The procedure for calculating an arithmetic mean is as follows:

1. Add all the data points to be averaged.
2. Divide the sum from step 1 by the number of data points. The resulting value is the arithmetic mean.

From Table 7.1, the total flow rate for the month was 689 (mil. gal). Dividing 689 by 30 days, the number of days, results in an average of 23.0 mgd (mgd $\times$ [3.785×10^3] = m^3/d).

WEIGHTED AVERAGE. A weighted average is a modified mean used with data points having different levels of importance (for example, the concentration of samples collected at different flow rates). A modified mean is calculated in the same manner as an arithmetic mean except that each value is multiplied by a weighting factor before being summed, and the resulting value is divided by the sum of the weighting factors. The weighting factor most commonly encountered is flow; however, other factors may be appropriate, depending on the data.

The procedure for calculating a weighted average is as follows:

1. Select a weighting factor for each data point and multiply the data point by the weighting factor.
2. Total the values generated in step 1.
3. Total the weighting factors and divide the sum from step 2 by this weighted average value. The resulting number is the weighted average.

The following example, which illustrates differing results from a arithmetic average and a weighted average, is based on Table 7.3.

The arithmetic average of BOD concentration values equals 172 mg/L. To determine the weighted average BOD concentration, the total BOD, 167 930 lb (lb × 0.453 6 = kg), is divided by both the total flow rate, 122 mil. gal (mil. gal × [3.785 × 10^3] = m^3), and the conversion factor, 8.34, resulting in 165 mg/L, which is less than 172 mg/L.

Keep in mind that DMRs may define average effluent BOD as the arithmetic average of the various concentrations and not the weighted average.

GEOMETRIC MEAN. A geometric mean is used to evaluate parameters that vary by much more than a factor of 10 or 20. The most common parameter is coliform count, which is expressed as a power of 10, such as 3.75 × 10^4/100 mL. The use of an arithmetic mean with such data is inaccurate because of its extreme variability.

Table 7.3   Weighted average example.

	Flow rate, mgd[a] Q	Concentration, mg/L BOD	Load, lb/d[b] = $Q \times$ BOD $\times$ 8.34
	15.2	195	24 720
	35.7	155	46 150
	22.2	175	32 400
	29.6	125	30 860
	19.3	210	33 800
Total	122.0	860	167 930
Average	24.4	172	33 590

[a] mgd × (4.383 × 10^{-2}) = m^3/s.

[b] lb/d × 0.453 6 = kg/d.

The procedure for calculating a geometric mean is as follows and uses Table 7.4 as an example:

1. Convert each data point to its logarithmic equivalent. Most hand-held calculators and spreadsheet software programs for computers have this function. Treat all zero values as 1.
2. Total the logarithms of all of the data points (12.902).
3. Divide this sum by the number of data points (12.902 + 9 = 1.434).
4. Take the antilog of the resulting number; this is the geometric mean. Thus, the geometric mean from Table 7.4 is 27.

An alternative method is to (1) multiply all data points in order (a × b × c...); and (2) take the Nth root of the product, where N = number of data points. For the preceding example, the product of the data points is 7.958×10^{12}. The 9th root of the product is 27, the same answer. For calculators that do not directly compute roots, the way to calculate the 9th root of the product is to take the product to the 1/9 (0.111 1) power, such as $(7.958 \times 10^{12})^{0.111\ 1} = 27$.

MOVING AVERAGE. Many parameters for evaluating and controlling biological systems have response times that extend several days or weeks. For example, it takes as long as a week for process changes to result in a change

Table 7.4   Geometric mean example.

Effluent fecal MPN[a]	(Step 1) Logarithmic value
2	0.301
350	2.544
23	1.362
5	0.699
2	0.301
79	1.898
240	2.380
79	1.898
33	1.519
(Step 2) Total	12.902
(Step 3) 12.902 divided by 9 equals 1.434	
(Step 4) The antilog of 1.434 equals 27	

[a] MPN = most probable number.

in the sludge-settling characteristics of an activated-sludge system. Thus, it is useful to look at a moving average when evaluating this type of data.

Moving averages can be constructed over any desired time interval; however, probably the most common is a 7-day moving average. An example of this calculation follows and is based on Table 7.5.

The 7-day flow rate moving average for day 13 (Table 7.5) is 22.8 mgd (mgd $\times$ [4.383 $\times$ 10^{-2}] = m^3/s). To find the average for day 14, the value for day 7 is dropped and the value for day 14 added. That sum is then divided by 7, leaving an answer of 22.5 mgd.

GRAPHING

Once the raw data have been reduced mathematically to a usable form, as in Tables 7.1 and 7.5, they must be presented in a manner that can be easily reviewed. The two most common approaches are to present data in either tabular or graphical form. Tables can be designed to any format; however, graphical presentations must conform to specified rules. Graphs normally consist of two axes at right angles to one another. The horizontal axis is called the x axis, and the vertical axis is called the y axis. Each axis is divided into intervals that correspond to the parameters in question. Graphs can be classified as linear, semilogarithmic, or logarithmic, depending on the manner in which the scales on the axis are divided. Each axis is associated with a specific parameter, and the parameters of the two axes are related on a one-to-one basis.

LINEAR GRAPHS. Linear graphs have linear or equal divisions on both axes. They are used to plot data that is linear or does not vary substantially. An example linear graph is Figure 7.1, which is a plot of average daily flow rate for each month. Figure 7.1 indicates that the wastewater flow rates at this particular treatment plant are higher during the winter months, probably in part because of infiltration and inflow.

SEMILOGARITHMIC GRAPHS. Semilogarithmic graphs are plotted on chart paper having one axis with a linear scale and another with a logarithmic scale. The logarithmic axis can be graduated on one or more orders of ten, referred to as *cycles*; thus, a graph having two orders of ten is called a *two-cycle* plot. Semilogarithmic plots are used when the data being plotted vary in a logarithmic manner; that is, when, plotted on a linear graph, they do not yield a straight line, or when one of the parameters being plotted varies over more than one order of magnitude. Figure 7.2 is an example of a semilogarithmic graph.

Table 7.5   Moving average example.

Date	Flow rate, mgd[a]	7-Day average	pH	7-Day average	Biochemical oxygen demand, mg/L	7-Day average	Total suspended solids, mg/L	7-Day average
1	23.6		7.2		272		182	
2	24.0		6.6		379		226	
3	24.4		6.6				192	
4	24.0		6.6		311		184	
5	23.8		6.7		291		175	
6	22.9		6.8		311		186	
7	22.0	23.5	6.8	6.8	226	298	168	188
8	22.9	23.4	7.2	6.8	274	299	154	184
9	24.1	23.4	6.8	6.8	371	297	148	172
10	23.5	23.3	6.7	6.8	358	306	184	171
11	25.0	23.5	6.8	6.8	302	305	176	170
12	22.4	23.2	6.7	6.8	336	311	196	173
13	19.9	22.8	6.7	6.8	296	309	168	171
14	20.0	22.5	6.6	6.8	248	312	153	168
15	21.2	22.3	7.0	6.8	286	314	192	174
16	21.4	21.9	6.7	6.7	326	307	193	180
17	25.5	22.2	6.9	6.8	332	304	190	181
18	25.1	22.2	6.7	6.8	322	307	184	182
19	21.1	22.0	6.6	6.7	306	302	168	178
20	20.5	22.1	6.8	6.8	313	305	174	179
21	18.3	21.9	7.0	6.8	245	304	186	184
22	24.0	22.3	7.2	6.8	234	297	162	180
23	25.2	22.8	6.7	6.8	320	296	176	177
24	25.5	22.8	6.9	6.9	290	290	175	175
25	24.2	22.7	7.2	6.9	258	281	166	172
26	21.0	22.7	7.1	7.0	207	267	150	170
27	21.5	22.8	7.2	7.0	199	250	142	165
28	22.6	23.4	7.0	7.0	206	245	158	161
29	24.2	23.5	7.2	7.0	225	244	160	161
30	25.1	23.4	7.0	7.1	263	235	182	162

[a] $mgd \times (4.383 \times 10^{-2}) = m^3/s$.

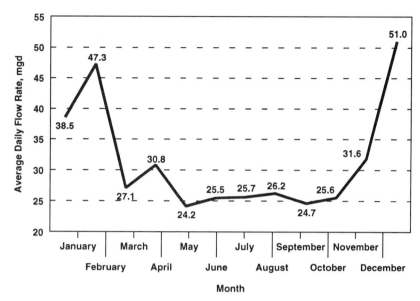

Figure 7.1   Linear graph example—average daily flow rate per month
$(\text{mgd} \times [3.785 \times 10^3] = \text{m}^3/\text{d})$.

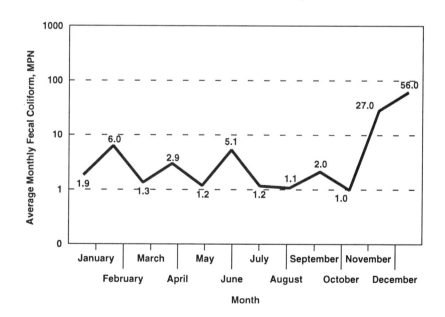

Figure 7.2   Semilogarithmic graph example—fecal coliform most probable number per month.

LOGARITHMIC GRAPHS. Logarithmic graphs are plots in which both axes have logarithmic scales. These graphs are used when both parameters being plotted vary widely or in a logarithmic manner.

G*RAPHICAL PRESENTATIONS*

Graphical presentations can be differentiated according to their various types. The most common types are bar graphs, trend charts, comparison graphs, and cumulative probability graphs.

BAR GRAPHS. Bar graphs typically are used to make comparisons between sampling sites, years, or other discrete units. The y axis is graduated in terms of the quantity being measured. The x axis parameter is not a discrete point but rather an area such as a solid bar. As a result, the plot is a series of rectangular bars, the heights of which correspond to the magnitude of the parameter in question. An example of a bar graph is presented in Figure 7.3, which is a plot of total BOD loading for each month.

TREND CHARTS. It is frequently valuable to plot or graph the variation of a specific parameter against time to isolate any trends. In trend charts, the x axis represents time; the parameter's magnitude, located on the straight line

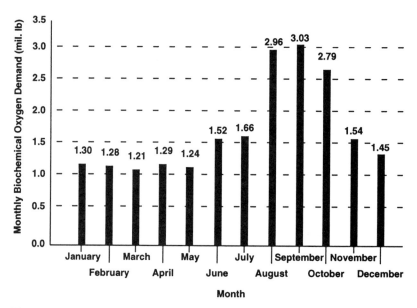

Figure 7.3   Bar graph example—monthly BOD influent loading (lb $\times$ 0.453 6 = kg).

---

drawn through the data, is generated by a mathematical method that will be described later.

SPECTROPHOTOMETRIC CALIBRATION CURVES. Another common graph is the spectrophotometric calibration curve, which compares percent transmittance with concentration. In this case, concentration is related to the logarithm of the transmittance, which leads to a logarithmic curve when plotted on a linear graph. However, this can be corrected for by using a semi-logarithmic plot, in which the transmittance is plotted on the logarithmic scale. As an alternative, absorbance (a logarithmic function of transmittance) can be plotted versus concentration on linear graph paper.

CUMULATIVE PROBABILITY GRAPHS. Cumulative probability graphs are plotted on special probability paper. The numbered axis represents the percentages of the values of the parameter measured that are equal to or less than a given value; the unnumbered axis can be graduated in the units of interest. The scale on this paper is designed so that data that fall in a normal random distribution will plot as a straight line. Once a cumulative probability plot has been constructed, it can be used to determine the probability of a parameter falling above (or below) any specified value. These data can be of great value in a number of circumstances, such as when evaluating the potential impact of a plant discharge on its receiving water or assessing the loading on the treatment plant.

A cumulative probability plot can be constructed from any quantity of data; however, the level of confidence that can be placed in the plot increases as the quantity of data increases. The procedure for constructing a probability plot is reasonably simple, although it can become time consuming when a large amount of data is being treated.

The steps are as follows:

1. Order all data in increasing numerical order.
2. Review the data to determine the range of values, and use this information to number the blank axis of the probability paper.
3. Determine the total number of data points.
4. At selected intervals, count the number of data points less than or equal to the selected value. Divide this number by the total number of data points and multiply by 100. This yields the percentage of the data that is less than or equal to the selected value.
5. Plot this data point on the probability paper. The point is located by drawing imaginary lines at right angles to each axis, starting at the values determined in step 4. The intersection of these two lines is the data point.
6. Steps 4 and 5 should be repeated until there are enough points for a line to be drawn. Depending on the nature of the data being plotted,

this line may be almost straight or, if the data do not fit a normal distribution, may be a curve.

PIE CHARTS. A pie chart depicts the relationship between components of data that add up to a total. An example is monthly rainfall for a year. This example would consist of 12 "slices" of pie, each slice representing the amount of rainfall in the month that it depicts. In other words, a slice representing a month that has substantial rain would be larger than that representing a month of little rain.

Figure 7.4 is an example of a pie chart that represents information similar to that in Figure 7.3. The quantity of BOD loading into a treatment plant is shown as a percentage of the annual total. In this particular example, the plant treats a large quantity of industrial food-processing waste during the summer months, as reflected in the chart by the August, September, and October BOD values being twice those of other months.

COMPARISON GRAPHS. A comparison graph contrasts one parameter with another. Each axis corresponds to one of the two parameters. This graphing technique normally assumes that the two parameters are related to each other in some way, the most common being a linear relationship. Figure 7.5 is an example of a comparison graph contrasting BOD and carbonaceous BOD (CBOD). This type of graph is also known as a *scatter-plot*, or *xy*, graph.

QUALITY CONTROL AND STATISTICAL EVALUATION

To ensure the accuracy of the analytical data being generated, it is important to evaluate and control the quality and reliability of the analytical testing pro-

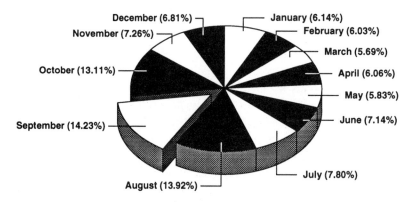

Figure 7.4   Pie chart example—BOD loading as percent of annual total.

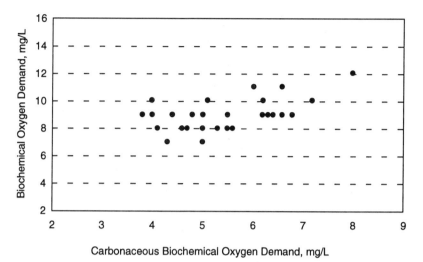

Figure 7.5 Comparison graph example—effluent BOD versus CBOD.

gram. A wide variety of considerations can affect the reliability of data, including sampling site, sampling technique, sample storage and transportation, analytical technique, and analytical calculations. Close control of all phases of the testing program (examined in detail in earlier chapters) is essential to ensure that the resulting data are of the highest quality possible. The techniques discussed in the following sections apply chiefly to the control and evaluation of the analytical testing program; however, some results may help point out areas of concern related to sampling and sample storage operations as well.

Although it is beyond the scope of this manual to develop an analytical quality control program, it is appropriate to outline the basic elements of such a program. U.S. EPA recommends that 10 to 20% of analyses be used for quality control purposes. U.S. EPA also recommends at that least one blank, one midpoint standard, one spike, and one set of duplicates be run with each batch of samples. Although it is common to refer to the *accuracy* of data, the quality of analytical data is determined by its *precision* as well.

ACCURACY. Accuracy is the first consideration in evaluating the quality of data. Accuracy is a measure of how closely the analytical result, or the average of a set of analytical tests, approaches the actual value of the parameter being measured. Ideally, they should be exactly the same; however, this goal is unlikely because all analytical tests are subject to a number of errors and limits that affect the accuracy of the final results.

Errors affecting analytical measurements can be classified as either systematic or random.

Systematic Errors. Also known as *cumulative* errors, systematic errors typically are the results of improper analytical techniques or inaccurate or faulty instruments (for example, inaccurately calibrated analytical balances). A systematic error, which causes a consistent error in the final result, can be eliminated by finding and correcting the cause of the error.

Random Errors. Also known as *accidental* errors, random errors cannot be prevented because some uncertainty exists in every physical measurement. Random errors are equally likely to cause positive or negative deviations from the true values. Examples of such errors are those that result during sample weighing and pipetting of samples, which, even when done as accurately as possible, presents an inherent inaccuracy associated with the class of pipet being used. It is possible to minimize the effect of random errors by performing a series of replicate analyses on the same sample and taking the average of these values. The overall reliability of the resulting average will increase as the number of measurements is increased. However, random errors limit the accuracy of the data. Under normal conditions, duplicate, or at most triplicate, analyses are used.

PRECISION. Precision is the second consideration in evaluating the quality of data. Precision is the reproducibility of a measurement or the consistency of a set of measurements (that is, the nearness of the individual measurements to each other). Precision is a reflection of how closely a series of replicate measurements approaches the average. It is typically assumed that the ability to produce data with high precision indicates that the results are also highly accurate; however, such is not necessarily the case. It is entirely possible to have excellent precision and poor accuracy or, conversely, excellent accuracy and poor precision.

Consequently, it is necessary to control both accuracy and precision to ensure that reliable data are produced. A number of methods are available for evaluating both accuracy and precision. These measures, however, do not take into consideration sampling and handling errors introduced before laboratory receipt of the sample.

Measurement of Accuracy. The accuracy of an analytical test is evaluated through the use of standard samples that have a known concentration of the constituent of interest. Standards for a number of parameters can be obtained either commercially, from U.S. EPA, or they can be made up from pure chemicals in the laboratory. These standards may be analyzed directly, or they may be added to unknown samples in a process known as *spiking*. Spiking is usually more desirable because it includes an evaluation of the effects of interfering substances on the sample being analyzed. The reader is referred to *Handbook for Analytical Quality Control in Water and Wastewater Laboratories* (U.S. EPA, 1979) for additional information on spiking.

PERCENT DEVIATION. Even though it is less dependable than spiked samples, direct analysis of standard samples is sometimes used to assess accuracy. The main problem with this technique is that it is preferable to work with a sample matrix as close to the unknown sample as possible. Standard samples rarely meet this criterion. The results of analyzing standard samples are expressed as percent deviations from the standard. U.S. EPA (1979) refers to this parameter as *percent bias*. They list standard values of this parameter for various analyses. As in all such procedures, the sample should be analyzed in replicate. Normally, seven replicate analyses of the standard are run to generate the data points for calculating percent deviation. The percent deviation is calculated using the following procedure:

1. Subtract the theoretical concentration from the average measured concentration as determined by the average of the replicate analyses.
2. Divide the value from step 1 by the theoretical concentration.
3. Multiply the result from step 2 by 100: this product is the percent deviation. It will be either negative or positive depending on whether the measured value was less than or greater than the theoretical value. The appropriate sign should be written as a part of the result.
4. The calculated value should be compared with the values published.

Example. Calculate percent deviation for the following analysis. A standard sample with a concentration of 15 mg/L is analyzed in triplicate. The results of these analyses are 14.7, 15.1, and 14.6 mg/L. The average of the three analyses is 14.8. The percent deviation is $(14.8 - 15.0)/15.0 \times 100\%$, or -1.3%.

SPIKED SAMPLES. The use of spiked samples, as indicated earlier, is preferable for evaluating the accuracy of an analytical procedure because it is performed on a sample that closely resembles the unknown sample. This procedure involves the addition of a known quantity of standard to a known volume of unknown sample. Replicate analyses of both the unknown and the spiked samples are run, and the results are compared to generate a percent recovery. Ideally, the result should be 100%. Typically, results from 90 to 110% ($\pm 10\%$) recovery are considered acceptable (see U.S. EPA, 1979).

MEASUREMENT OF PRECISION. Precision is evaluated by analyzing a set of replicate samples. The simplest way to measure precision is to determine the range of a series of replicate samples. This method can be used most effectively on small sample sizes. Several additional measures of precision are used, including average deviation, variance, and standard deviation. The latter two parameters are typically used for the statistical evaluation of analytical data.

AVERAGE DEVIATION. The procedure for calculating average deviation is as follows:

1. Calculate the average for the set of data.
2. Calculate the difference between each data point and the average; treat all values as positive, even if they are negative.
3. Total the values calculated in step 2.
4. Divide the results from step 3 by the number of data points. The result is the average deviation.

Example. Calculate the average deviation of the following data set. The results of a series of analyses of an unknown sample are as follows: 14.7, 15.3, 15.1, 14.2, 14.9, and 15.0. The average value is 14.87. The positive differences between the average and the individual values are 0.17, 0.43, 0.23, 0.67, 0.03, and 0.13. The sum of the values from step 2 is 1.66. The result from step 3, divided by the number of data points, is $1.66/6 = 0.28$. This is the average deviation of the data set.

VARIANCE. The variance, although not a direct measurement of precision, is closely related to the standard deviation and used extensively in statistical calculations. There are a number of methods for calculating the variance, all of which are time consuming. Many calculators and computer programs are preprogrammed for this capability; thus, calculating the variance may be as simple as entering the data and pressing the appropriate keys.

Otherwise, the following simplified approach can be used:

1. List each value and calculate its square.
2. Total these values.
3. Total the individual values, square the result, and divide by the number of samples.
4. Subtract the value in step 3 from the value from step 2.
5. Divide the result by $n - 1$ (one less than the number of samples); the resulting value is the variance.

STANDARD DEVIATION. The standard deviation is the most commonly used estimate of precision for evaluating moderately sized sample populations. The standard deviation (s) is defined as the square root of the variance. The procedure for calculating the standard deviation thus involves determining the square root of the variance. As will be discussed later, it is possible to estimate the confidence interval of analytical data once the standard deviation has been determined. Standard deviation is also a useful parameter for rejecting outlying data points.

Most calculators and computer spreadsheet programs can automatically calculate standard deviation.

QUALITY CONTROL CHARTS. Quality control charts provide a means of evaluating day-to-day performance, once sufficient accuracy or precision data have been generated. Accuracy data are generated through the analysis of standard and spiked samples; precision data come from the analysis of duplicate samples. Normally, 15 to 20 sets of data, either duplicates or duplicate spikes, are required to provide sufficient data for constructing a quality control chart.

The three types of quality control charts considered include cumulative-summation, precision, and accuracy charts. In all cases, the charts consist of an *x* axis, which corresponds to the time or order of the results, and a *y* axis, which is graduated in the units of the test value. The *y* axis includes an average or control value, an upper and lower control limit, and, in the case of precision and accuracy data, an upper and lower warning limit. The limits on these charts are calculated from actual analytical results coupled with decisions concerning the acceptable level of variability in the data and the degree to which a technician is willing to accept data that are not within control limits or reject data that are within control limits. A brief discussion of several of these charts follows (see U.S. EPA, 1979).

Cumulative Summation. A cumulative-summation graph plots the sum of the sample variance versus the sample set number. Duplicate analyses of samples and standards are run initially to attain the necessary data to generate the chart. Analyses are then run at selected intervals to verify that the test is still within control limits. The selected interval between duplicates should ensure that an excessive period of time does not pass before it is discovered that the test is outside of control limits. At the same time, the selected interval should not allow excessive analytical work load.

Three basic conditions can result from cumulative-summation charts: results can be within control limits, greater than the upper limits, or less than the lower limit. The procedure to follow in each case is as follows:

1. *Results within control limits*—no problem exists; therefore, continue with the analyses.
2. *Results greater than upper limits*—stop the analyses. Determine the cause of the deviation from acceptable performance and correct the problem. Repeat the analyses of samples run between the last test within control limits and the test outside of control limits. Start a new chart beginning at the first setpoint.

    This approach should be followed whenever the most recent results exceed the upper limit or when an established trend indicates that this limit ultimately will be exceeded. If precision data are being analyzed, such errors can be attributed to the analyst, the type of sample, or contaminated glassware. If accuracy data are being analyzed, errors can be attributed to the analyst, contaminated glassware or reagents, instrumentation problems, or sample interference on spiked samples.

3. *Results less than the lower limit*—this indicates either an improvement in the test performance or false data reporting. Continue the analyses and develop sufficient data to construct a new chart.

Precision Control Charts. Another quality control chart is the precision control chart. Precision control charts combine the data range and a set of factors that are based on an assumed level of confidence. They are used to calculate both an upper and lower warning limit and an upper and lower control limit. As the limits' names imply, values greater than the warning limit but less than the control limit indicate a condition that, though not necessarily outside of control limits, is cause for concern. Based on the assumptions used for constructing the precision control chart, approximately 5 out of every 100 sets of data should be greater than the warning limit. A test can be considered outside of control limits when substantially more than 5% of the data are greater than the warning limit and when the data points exceed the control limits. In addition, any trend of data toward the control limits is disturbing. All procedure variables should be checked when such a trend is observed to ensure that the limits are not exceeded. Once the control limit has been exceeded, all testing should be halted until the cause has been determined and corrected. After the test is back within control limits, all analyses run between the last acceptable set and the out-of-control set must be repeated.

Accuracy Control Charts. Accuracy control charts are similar to precision control charts except that they are constructed using data collected on spiked samples. Range data from the spiked samples are used to construct both precision control charts and accuracy control charts (refer to U.S. EPA [1979] for the procedure for constructing accuracy control charts). Interpretation of accuracy control charts is the same as for precision control charts.

LINEAR REGRESSION. A number of cases exist in which two parameters are related to each other in a linear manner. One of the most common of such relationships involves spectrophotometric analyses and the correlation of sample concentration to sample absorbance. Because of the frequency with which linear relationships are encountered, it is useful to have a technique for evaluating how closely two parameters approach a linear relationship. In addition, one should be able to construct the best straight line through a set of data when a linear relationship is suspected.

Such a procedure exists and is known as a *linear regression*, or *least-squares-best-fit*, analysis. This procedure can be used to generate a linear equation of the form $y = mx + b$, where

x = the individual value of the x parameter;
y = the value of y, which corresponds to the x parameter;
m = the slope of the line, y/x; and
b = the point at which the line intersects the y axis, which is known as the y *intercept*.

This approach treats one of the two variables as an independent variable and the other as a dependent variable. Thus, y is assumed to be dependent on the value of x. Using the same data as were used to generate the equation for the straight line, it is possible to generate the correlation coefficient r. This parameter is an estimate of how close the data actually come to fitting a linear relationship. Calculators are programmed to perform a linear regression and calculate r. This same procedure can be done manually, though it is relatively time consuming. If such a calculation is to be performed often, it is probably more efficient to purchase a calculator with this capability. Once developed, the equation can be used to calculate any value of y that corresponds to a selected value of x. This shortcut is especially useful with standard curves for which the concentration can be calculated directly from the derived equation rather than from a visual examination of the curve.

The value of r can range from -1 to $+1$, depending on whether the line has a positive or negative slope. A 0 value indicates that there is absolutely no linear correlation, whereas a ± 1 value indicates perfect correlation. Typically, r will lie somewhere between 0 and ± 1. The probability that a linear correlation actually exists becomes greater as the value of r approaches ± 1. Table 7.6 indicates the approximate values of r at various sample sizes for which there is at least a 95% probability of a linear relationship.

REJECTION OF OUTLYING DATA. Ideal analytical procedure requires that multiple analyses be performed on each sample and that the average value of the set be reported. This procedure provides the most accurate measure of the sample value by averaging out the random errors associated with

Table 7.6   The correlation coefficient (r).

Sample size	Minimum r for 95% probability of linear relationship
4	±0.96
5	±0.88
6	±0.82
7	±0.76
8	±0.71
10	±0.64

any analytical technique. At the same time, the probability of obtaining an out-of-limits value increases as the number of analyses increases. As a result, it is desirable to have an objective means for rejecting data that are suspected of being incorrect. Several such methods based on precision measurements have already been described. The three parameters discussed below are the average deviation, standard deviation, and Q test. The average deviation is the least rigorous statistically. Under normal conditions, the Q test is the most appropriate when the sample size is 3 to 10, whereas the standard deviation is best when the sample size exceeds 10.

Average Deviation. When the value obtained is questionable, calculate the average deviation of the data set with the questionable data point excluded. If the difference between the average value and the questionable value is more than four times the average deviation, then the questionable value can be rejected. If the value is rejected, the average must be recalculated. This procedure is not statistically rigorous; however, it is highly conservative and will reject only grossly variant data. The collection of additional data may result in a smaller average deviation, thus allowing the questionable value to be rejected.

This procedure should not be used with sample sets having fewer than four data points.

Standard Deviation. Once sufficient data have been collected on an analytical procedure to allow the accurate calculation of its standard deviation, this parameter can be used to assess outlying data for rejection. If standard deviation data are unavailable and sufficient data have been collected in the data set, a standard deviation excluding the questionable value can be calculated. If the difference between the average value of the set and the questionable value exceeds three times the standard deviation, it can be rejected with confidence of greater than 95%.

Although statistically valid, this procedure is of questionable reliability when the sample size is less than 10.

Q Test. For data sets having 3 to 10 data points, a parameter calculated using the range and the questionable data point is a more valid measure for rejecting the data point than either the average or standard deviation. This parameter (Q) is calculated by dividing the range of the data set into the difference between the questionable value and the value closest to it. If the calculated Q value exceeds the value listed in Table 7.7 for the appropriate value of n, then the questionable value can be rejected with 90% confidence that it is significantly different from the rest of the data points. The test can be repeated on the remaining data until all questionable data points have been rejected. This is an iterative process.

Table 7.7   Values of Q at 90% confidence level.

n	Q at 90% confidence
3	0.94
4	0.76
5	0.64
6	0.56
7	0.51
8	0.47
9	0.44
10	0.41

STATISTICS FOR NONDETECT DATA SETS. Data sets of chemical concentrations less than the analytical chemistry limit of detection (LOD) value are becoming increasingly more common as permit limit conditions increasingly include monitoring requirements and limitations for trace organic contaminants. Observations quantified as less than LOD, or nondetect (ND), result when the compound being measured is either absent or exists at such low concentrations that laboratory personnel judge the analytical result unreliable. This data censoring by laboratory personnel can take place for a variety of reasons, including noncompliance with quality control/quality assurance criteria such as slow signal-to-noise ratio or poor recovery of spiked compounds.

The presence of less than LOD or ND observations in a data set makes the estimation of summary statistics more involved: the lack of quantitation associated with the censored observations prohibits their direct use in simple straightforward calculations of common statistics such as arithmetic mean. For most environmental monitoring situations, the censored observations represent random values between the LOD and 0. Although the substitution of a fixed value for censored observations followed by common statistical calculations is an expedient (and common) way of converting a cumbersome analysis into a simple one, such an approach yields biased estimates and is not recommended because other statistical methods with only modest computational burdens are available. Statistical method bias can be particularly important when the decrease in an environmental variable is being monitored because the degree of bias increases with the number of censored observations in the data set. This results in a distorted picture of the environmental variable's change.

A consensus approach for handling ND observations does not presently exist. The interested reader is directed to Statistical Procedures for Addressing Non-Detect Results in Environmental Analysis (Hinton, 1994), which provides an overview of sensible alternatives for handling ND observations, including general analysis procedures for censored data sets and three

---

commonly used methods for estimating means and standard deviations from normally distributed data sets that are singularly left censored.

COMPUTER SOFTWARE AVAILABLE

There are numerous desk-top computer "spreadsheet" programs available today at reasonable prices. There are also database and statistical programs available with additional features.

Most spreadsheet programs have a matrix of cells or spaces into which data are entered. They typically run across the page (rows) and down the page (columns). Tabular data are commonly entered down the column for a particular parameter, such as flow, BOD, or temperature. The user can perform mathematical functions with the data, with the results automatically calculated and entered into an adjacent column.

The options for displaying the tabular data are almost unlimited with modern spreadsheet programs. Titles, borders, margins, footnotes, headers, shading, and various font sizes are available.

Probably the most common method for selecting software is testing it out on an existing system. Most software programs can easily perform simple functions such as addition and multiplication. However, the user must have patience and time to master the wide range of functions available, such as statistical analysis and complicated graphics. Most users eventually use only a small fraction of a particular software's capabilities, but the full range of analysis is always available.

Numerous hand-held calculators with preprogrammed statistical functions are readily and economically available through business stores and catalogs. For simple statistical analyses that do not need to be repeated, calculators are convenient tools.

REFERENCES

Hinton, S. (1994) Statistical Procedures to Addressing Non-Detect Results in Environmental Analysis. *Tappi J.*, April.

U.S. Environmental Protection Agency (1979) *Handbook for Analytical Quality Control in Water and Wastewater Laboratories.* EPA-600/4-79-019, Cincinnati, Ohio.

U.S. Environmental Protection Agency (1988) *NPDES (National Pollutant Discharge Elimination System) Compliance Inspection Manual.* PB 88-221098.

Chapter 8
Data Reporting

INTRODUCTION

Data reporting is the final step in a quality wastewater sampling program. The data collected through sampling and analysis become useful when reported in a concise and logical manner. A good plant sampling program may produce representative and quality data, but unless the data are handled properly and reported to the appropriate agencies or people, they may end up being unusable information. Therefore, proper reporting is, in fact, as important as the analysis itself.

Often, samples of a process stream are collected and analyzed only to be stored and forgotten in a workbook. At other times, a particular piece of data may be meaningless without additional information. For example, a solids stream is analyzed for the concentration of solids, but the flow records on the strip chart are in a remote or obscure location. Without the information on this chart, the plant manager may be unable to determine why there is a change in solids concentration.

Another common problem with wastewater sampling programs is the reporting of numbers that, at some point, have been copied or calculated wrong. The more numbers a person handles, the greater the chances are of making mistakes. Errors in handling numbers can, in an extreme case, cause a

---

reported violation of the plant's discharge permit or lead to the misapplication of a chemical.

Preprinted bench sheets should be used to reduce errors in handling data and increase the usefulness of the data generated from the laboratory. A well-prepared form also will reduce the amount of time spent recording information. Appropriate spaces can be made available for recording data in the right order, and the steps and calculation procedures can be outlined.

Workbooks should be used to store all of the bench sheets for a given month, along with the daily summaries. This keeps all of a month's data in one location and eases the task of completing the monthly reports required by the regulatory agencies.

Special consideration should be given to the purpose of a report. The report format should be designed to be reader friendly so that it can benefit the party that requires the data. For example, the regulatory agency is primarily concerned with the quality of the effluent discharged. The city clerk, on the other hand, may need to know the mass of a particular substance to process a billing statement. Other possible uses of data may include reports to the city or another governing body, public communications, and historical documentation. Understanding the needs of each reader of these reports will reduce follow-up questions and explanations.

*P*ROPER DATA HANDLING

Proper data handling requires an efficient manner of distributing information including considerations of method, speed, and accuracy. Facility operations become more efficient when a method has been established to transmit information from the laboratory to the appropriate personnel. Bench sheets should be designed to simplify the retrieval of key information and must be filed systematically so that data can be located easily. The use of workbooks and daily summaries are both useful methods for filing data.

To speed up process control decisions, industrial waste enforcement actions, regulatory reports, and other procedures important to daily plant operations, data must be transmitted as quickly as possible. As soon as information becomes available, the relative value of the data must be weighed against its intended use. If process control tests indicate abnormal conditions, the data must be transmitted as soon as possible, usually verbally, and followed by written documentation. Compliance monitoring data can be tabulated as the numbers come from the laboratory. Such data are then left until the end of the month when minimums, maximums, and averages can be calculated.

BENCH SHEETS. Filling out the bench sheet is the first step of data handling in the laboratory and the most important. Bench sheets should be preprinted forms that allow personnel to record data easily and retrieve them easily when necessary. The arrangement of the information on the bench sheet determines the ease of data recording and retrieval.

The preprinted form prompts the analyst on which data to record and typically indicates the number and type of samples to be analyzed along with which tests are to be run. A good bench sheet provides space for the necessary calculations and formula and for all quality assurance and quality control (QA/QC) data. Thus, all data related to a given test can be recorded easily in the same place.

To ensure the proper use and quality of data in bench sheets, it is important to record data permanently in ink and initial and date all changes to data previously entered, express data in the same units and number of significant figures, store data for easy retrieval, and document QA/QC data.

Pencils should not be used on bench sheets. All data should be recorded in ink, preferably indelible ink. All data should be written clearly: a number that is difficult to read lends itself to errors. A bench sheet lying on the top of a laboratory bench is susceptible to spills or splashes; the use of a clip board will help prevent this problem. If a recording error is made, it is not acceptable to erase it. The appropriate method of correcting a recording error is to draw a line through the mistake, record the proper number next to the erroneous data, and initial it.

All bench sheets should use the same units of expression. If, for instance, plant loading is required in kilograms per day (kg/d [lb/d]) or an industrial source as population equivalents, these units must be expressed. The most common way to report inorganic test data is in milligrams per litre (mg/L), while organics are reported in micrograms per litre (μg/L). These units can be converted readily to other units by the person who has requested the data. It is important to establish how the data will be used and report values in the most appropriate units. The fewer times a value has to be plugged into a formula, the less likely it is that an error will be made.

Another matter of concern is the number of significant figures (Chapter 7) represented in the data. As a general rule, all factors in an equation should be expressed to the number of decimal places in which the analyst is confident. The reported result then should be rounded off to as few significant figures as are present in the factor with the fewest significant figures. (Typical wastewater laboratory data are expressed in two, or at the most three, significant figures.)

In addition, data storage should be well planned. Bench sheets and calculations should be filed so that the information can be retrieved readily. Cross-references in the filing system between the bench sheets, workbooks, daily summaries, and resulting reports are helpful. Today, many wastewater treatment plants use computer-based systems to store laboratory data.

All QA/QC data generated during the development of data required for reporting purposes should be available for easy reference. Stating the precision and accuracy provides the person evaluating the reported value with reliability in using the laboratory information. This person might have more confidence in a value reported as 100 ± 1 than in a value reported as 10 ± 1. If triplicate analyses are run, ranges, means, and standard deviation values may be reported.

Bench Sheet Design. Laboratory bench sheets in a treatment plant can help provide a solid foundation for data flow and also serve as final checks in all questions of error and credibility. It is important for data to be both credible and accurate. The laboratory data should answer the following questions:

- What samples and analyses were run?
- Is the analytical procedure working properly?
- Was the sample noticeably different or unusual?

The information collected with each test varies, but in all cases, there should be enough data not only to obtain final results but also to control the test in question. For example, a bench sheet of a titration should include more than the volumes of sample and titrant used. Provisions should also be made to include standard and duplicate samples. Thus, changes in titrant are corrected promptly and also documented. This additional documentation significantly increases laboratory credibility, a prime ingredient in any self-monitoring program.

A more complex example might be a bench sheet for the membrane filter coliform test. In addition to sample volume and final colony count, various other information should be collected. Blanks should be run before and after sample filtration to demonstrate the sterilization of equipment and adequacy of rinsing techniques. In addition, the temperature of the incubator should be measured and the lot number and manufacturer of the media and date the media are prepared should be recorded. Other methods of quality control can also be applied. In all cases, this information should be recorded on the bench sheet with the data needed to determine the actual results.

Finally, an important part of a laboratory bench sheet that is often forgotten is the "comments" column. The comments can account for some of the most important data collected. Here, both the operator and analyst play a role. Anything unusual noticed by either should be placed in the comments column. Such observations should include

- Smell—does the sample have a characteristic odor? For example, does it smell like beer, fruit, oil, or septicity?
- Color—is the solution colored? Color could indicate something as innocuous as the presence of food coloring or as severe as a heavy metals discharge.
- Appearance—does the sample appear visually different? For example, is it excessively gritty, or can the presence of grain or oil, for example, be detected?

- Other—have there been major process changes or upsets? Is there any problem during testing?

This information typically is unnecessary for the laboratory test, but it becomes critical in the interpretation of the test, particularly when abnormal results are analyzed.

Whenever possible, for reasons of credibility, it is wise to fill the comments section out either before or during the testing procedure. Typical bench sheets are shown in Figures 8.1, 8.2, and 8.3.

DAILY SUMMARIES. The second step in facilitating the flow of information within wastewater treatment plants is filling out the daily summary. Daily summaries reflect the operation of a specific facility and include a variety of data regarding the laboratory, process control, collection system, pumping stations, or industrial waste contributors. The purpose of the daily summary is to make the data generated in the facilities available to operators, engineers, and planners.

A number of steps should be followed in setting up a daily summary:

- Decide what information is needed, required by, or of interest to the people reading the summary.
- For purposes of review, have one person complete the summary sheet and another review it for correctness and examine abnormalities (it is in this capacity that bench sheet comments are important).
- To minimize the possible misinterpretation of data by nonlaboratory personnel, establish a key to laboratory results.
- Organize summary formats. Often, more than one daily summary can be generated. For example, two separate summaries might be made, one for solids handling and the other for liquid processes.

Any daily summary should include the date the sample is tested, sample source and type, required data and units (including accuracy or range where appropriate), and signature of reviewer. The emphasis should be on getting correct information to those who need it in a timely fashion. A typical daily summary worksheet is shown in Figure 8.4.

WORKBOOKS. The third step in data handling is the workbook, a compilation of all data collected. In recent years, the U.S. Environmental Protection Agency (U.S. EPA) has encouraged the use of bound workbooks (laboratory logs) in which to store all laboratory information as part of a QA/QC program. Information would include such items as instrument calibration, oven and incubation temperatures, and the use of duplicate and split samples. An expansion on U.S. EPA's version of the workbook would be a workbook for each month of the year that includes all laboratory data except for annual reports.

Sample Date: _____

| Collected By: _____ Time Collected: _____ | Date In: _____ | Date Out: _____ Weekday:_____ |
| BOD Set Up By: _____ Time Set Up: _____ | BOD Read By: _____ | Time Read: _____ |

Bottle	Size	Sample	DO Initial	DO Final	Depletion		BOD
		Initial Blank					
		Final Blank					
	5.0 ml.	Raw Inf.					
	7.0 ml	Raw Inf.					
	9.0 ml	Raw Inf.					
	6.0 ml.	Pri. Eff.					
	8.0 ml.	Pri. Eff.					
	10 ml.	Pri. Eff.					
	60 ml.	Sec. Eff.					
	80 ml.	Sec. Eff.					
	100 ml.	Sec. Eff.					
	60 ml.	Sec. Eff. Dup.					
	100 ml.	Sec. Eff. Dup.					
	60 ml.	Sec. Eff. Spike					
	100 ml.	Sec. Eff. Spike					
	100 ml.	Sec. Eff. N. Inhib.					
	15 ml.	Pri. Eff. Soluble					
	25 ml.	Pri. Eff. Soluble					
	5 ml.	ML. Inf. Soluble					
	25 ml.	Pri. Eff. Soluble					
	60 ml.	ML. Eff. Soluble					
	100 ml.	ML. Eff. Soluble					
	100 ml.	2nd Day BOD					
		DAY OUT =	Ratio =				

EPA Method 405.1 24 hours Holding Time on BOD 5
BOD Spike on Even Num. Day   BOD Duplicates on Odd Num. Days

| % Reduction (Inf to Eff) = _____ |
| % Recovery Spike = _____ |

Figure 8.1   Typical bench sheet, example 1.

A three-ring binder would be easiest to use, though U.S. EPA encourages the use of bound workbooks to ensure that the data have not been tampered with. Therefore, it is recommended that a three-ring binder be used for the current month and that all information be bound and filed at the end of each month. Each monthly workbook could include the original bench sheets, all

DAY _____

DATE _____

OPERATOR _____

	TIME		TIME				
	TEMP (°F)	pH	TEMP (°F)	pH	AMMONIA	SETTLEABLE SOLIDS	TURBIDITY
Raw							
Primary							
Secondary							
Final							

SUSPENDED SOLIDS

	Raw	Primary	Secondary	Final	Grab Flow			
					Mixed Liquor		Return Sludge	
					pH	pH		MLVSS
Filter No.								
Sample Size (ml)								
Paper & Dry (g)								
Paper								
Dry Weight (g)								
% Removal								

Sample Time	Flow	Composite Sample Size	PPM Residual	D.O.P.P.M.		TOTAL COMP.
				Time	FLOW	
				Pass 1	FINIS #	
				Pass 2	START	
				Pass 3	TOT (MS)	
				Pass 4		
				Final		
				Wasting (W.A.S.)		
				Gal. Time		

COMMENTS: _____
_____
_____
_____
_____
_____

Figure 8.2   Typical bench sheet, example 2 (gal $\times$ [3.785 $\times$ 10^{-3}] = m^3).

QA/QC information, the daily summaries, and monthly regulatory reports. The workbook becomes a powerful tool providing a single source for storing all laboratory data. From the notebook stage, personnel can move directly and efficiently to report writing.

USA WASTEWATER TREATMENT FACILITY
BACTERIOLOGICAL ANALYSIS WORKSHEET

DATE: _____

LOCATION OF SAMPLE COLLECTION: _____

Sample Collected By: _____   Time: _____

Analysis Performed By: _____   Time: _____

Plate Readings by: _____   Time: _____   Date: _____

Chlorine Residual: _____

Precontrol: _____   Postcontrol: _____

E. COLI

SAMPLE SIZE	# OF COLONIES	#/100 ML.

E. Coli = _____

TOTAL COLIFORM

T. Coliform = _____

Confirmation of Total Coliform:

Trypticase Glucose: (positive or negative)

#1   #2   #3   #4   #5

_____   _____   _____   _____   _____

Set up by: _____   Date: _____   Time: _____
Read by: _____   Date: _____   Time: _____

Brilliant Green: (positive or negative)

#1   #2   #3   #4   #5

_____   _____   _____   _____   _____

Set up by: _____   Date: _____   Time: _____
Read by: _____   Date: _____   Time: _____

Holding Time: 1 hour

Method: E. Coli, EPA 1103.1 Membrane Filtration

       T. Coliform, Std. Methods 908A

Figure 8.3   Typical bench sheet, example 3.

DATE: _____ _____

| Secondary Effluent Flows | Total:_____ | Minimum: _____ | Maximum:_____ |

RAW INFLUENT

Collection Time: _____

Sampled By: _____

Time in Lab: _____

Analyst: _____

Analyte	Results	Analysis Time	Duplicate
pH (std. units)			
S/S ml/L			
Temp. C			
D.O. mg/L			
Sulfides mg/L			

SECONDARY EFFLUENT

Collection Time: _____

Sampled By: _____

Time in Lab: _____

Analyst: _____

Analyte	Results	Analysis Time	Duplicate
pH (std. units)			
S/S ml/L			
Temp. C			
D.O. mg/L			
Sulfides mg/L			

CHLORINE RESIDUAL   (mg/L)   **Secondary Effluent**

Results	Collect Time	Sampler	Time in Lab	Analysis Time	Analyst

Cl. Standard: _____ Absorbance: _____ Absorbance OK: _____ Analyst:_____ Time: _____

Methods: pH - EPA 150.1, Settleable Solids - EPA 160.5, Temp - EPA 170.1, D.O. - Std. Methods 17th 4500 OG, Chlorine Residual - EPA 330.5, Sulfides - Hach Total Sulfides w/DR2000, All samples are grabs.
Holding Times: pH - 15 minutes, Settleable Solids - 30 minutes, Temperature - None, D.O. - 15 minutes, Chlorine Residual - 15 minutes.
Windows: Daily, Pub.
Reviewed By: _____ Date: _____

Figure 8.4   Daily summary worksheet.

REPORT PREPARATION

Typically, all data collected at a wastewater treatment plant ultimately end up in a report directed to some person other than the one who has collected the data. Each report compiled from plant data should be designed to provide specific information in a standardized reporting format. This format should compare the analytical or raw data to some kind of baseline data that has been established for the area of interest. For reports to governing boards on plant loadings, for example, plant design capacities should be reported as well as plant loadings. The plant design numbers give the reader a reference point for reviewing the data on plant loadings.

A report to the regulatory agencies must include all of the required information in a format that permits the data to be compared with effluent limitations set forth by that particular agency. Reports to be made public must be concise so that misinterpretation of the data is unlikely. To establish base charges and surcharges, loading limitations and actual loadings should be reported to industrial dischargers. In enforcement actions, actual measured violations should also refer to the legal justification establishing the violated limits.

It is advisable to keep process control data for an indefinite period of time because the data become useful in determining changes to the treatment process. Process control data can be analyzed using trend charts developed over a long period of time.

Information, particularly historical data, that can be used to upgrade certain process units should be maintained indefinitely. All data developed during studies of particular process units also should be kept according to regulatory and owner policies.

There are at least six basic steps in preparing a report:

1.   Defining objectives,
2.   Establishing fixed times and reporting frequencies,
3.   Compiling data,
4.   Designing formats,
5.   Reducing data quantity, and
6.   Summarizing.

The author needs to identify a reason for writing the report and then report the data in a way that will support a certain conclusion. Admittedly, the original conclusion may be altered as the report progresses: examining and ordering data are the chief ways of reaching a conclusion.

Typically, reports focus on data within a specific time period. If, for example, the report objective is to determine the food-to-microorganism (F:M) value of the activated-sludge system, a time period must be defined. An F:M value for one week's worth of data could have an implication quite different from an F:M value for a year's worth of data. To minimize future work, the frequency of reporting these objectives should also be established. If the F:M value is to be reported monthly, then more attention should be given to setting up data and calculation sheets.

The next step is to determine data needs. Are more data needed? Is the laboratory equipped or staffed to provide more data? Do other people have data that will be needed in the report? Answering these questions will be helpful in avoiding unnecessary work. When data needs are determined, it will be possible to estimate when the report can be finished. It may be that there is not enough time to finish a report to meet a particular need. It is better to make such a determination as soon as possible.

The fourth step, report design, should be made easy by following the first three steps. The objectives are fairly clear, and the quantity and type of data needed have been made available. The report format must be logical and address the audience that will read the report.

Data reduction can be viewed as a check point. The data have been collected and the data columns completed. Now the task is to present only the important data. Would it be helpful to give maximums, minimums, and averages? Should data from past reports be referenced? Should the data be shown on a graph? The answers to these questions will help the technician decide how to present the data so that it best carries out the objectives.

The final step is to provide a written summary, if one is needed. The summary should consider whether conclusions or recommendations are needed. Do the conclusions require some qualifications? Maybe the data, such as carbonaceous biochemical oxygen demand values, were determined only during the summer months. In that case, the data are likely to be inapplicable to the winter months. Giving the conclusions first lets the reader decide whether or not to look at the detailed data. Often, the reader may never do so, unless the report indicates particular problems.

Reports should be typewritten or computer generated except when using preprinted report forms with standard information. Each report should clearly state the purpose of the report and contain appropriate signatures, dates of information, and the distribution list.

Periodically, all reports and laboratory work should be reviewed and a determination made as to what should be dead filed (work that has been completed or will never be used).

LABORATORY REPORTS. When analyses have been completed, the bench sheet is reviewed by the laboratory supervisor or the person responsible for reviewing the bench sheet. This person should cross-check all calculations and all other laboratory data. Any extreme values in the analyses, such as extremely high or low oxygen depletion in a biochemical oxygen demand (BOD) test, should be noted and investigated. Other abnormalities must be noted and reported according to the nature of the abnormality. Data falling outside of process control limits must be reported without delay. Results outside of normal ranges or exceeding effluent limitations may be tied to upsets in various process units; therefore, it is necessary that the report indicate these values immediately. In addition, routine inspection of data is essential to the recognizing of trends as they develop. The person responsible for the facility should sign all laboratory reports that go to outside agencies.

The daily laboratory results should be given to the person in charge of operations, as well as to the plant manager. The chief operator should keep copies of reports containing process control data. Industrial waste inspectors also can be given copies of the laboratory reports containing surveillance data.

The plant manager should see this report as well. Originals of laboratory bench sheets and the workbook should never leave the laboratory.

It is essential that the laboratory data be made available to the plant operators (refer to Figure 8.5). Operations personnel should be encouraged to use the data to operate the plant more efficiently. As they become aware of problems in the plant, the operators should notify the lab so that a decision can be made on whether to take additional samples. Both training and in-plant experience enable an operator to identify a trouble spot and recommend a change in a process before the problem gets totally out of control.

USA WWTF
OPERATIONS DATA SHEET

Date: _____

Sample Date: _____

Flow (to aeration)............................... _____ MGD
Influent Suspended Solids .... _____ mg/l
Primary Influent Suspended Solids............. _____ mg/l
Primary Effluent Suspended Solids .... _____ mg/l
Final Effluent Suspended Solids.............. _____ mg/l
Final Effluent Suspended Solids
      Monthly Average.................... _____ mg/l

Sample Date: _____

Mixed Liquor Suspended Solids .... _____ mg/l
R.A.S. Suspended Solids............................ _____ mg/l
Mixed Liquor Settleable Solids .... _____ mg/l
Respiration Rate:
      Mixed Liquor.................... _____ mg/g/hr
      R.A.S............................ _____ mg/g/hr

FIVE DAY MOVING AVERAGE (average of last five values)

Influent Suspended Solids .... _____ mg/l
Final Effluent Suspended Solids.................... _____ mg/l
Primary Effluent B.O.D............................... _____ mg/l
Mixed Liquor Suspended Solids.................... _____ mg/l
R.A.S. Suspended Solids............................ _____ mg/l
Flow to Aeration....................................... _____ mg/l

Primary Effluent B.O.D.:

Date: _____   _____ mg/l

%Total Solids:
Date: _____
Feed: ....
Pressate: .... _____
Filter Cake: .... _____

Primary Influent pH................................... _____ s.u.
Final Effluent B.O.D. Monthly Average.......... _____ mg/l
Fecal Coliform Monthly Average................... _____ MPN/100ml

          Signed   _____

Figure 8.5 Operations data sheet (mgd $\times$ [3.785 $\times$ 10^3] = m^3/d).

If efficiently reported, the data generated by the laboratory can be used to control the treatment plant process effectively.

PLANT OPERATIONS. Reports to plant management should correlate with process data or previous reports. This correlation is made by comparing existing data to the new information. A reference to a previous report containing specific existing data may be sufficient if only some of the new data are going to be reported.

The most important plant report is the monthly report, which serves several purposes. It can be the official record of plant performance to regulatory agencies and management (Figure 8.6). The monthly summary also can be used to review plant operations and develop process control strategies (Figure 8.7). Finally, the monthly report is used as a data source in long-term planning. Data included in a monthly report are defined by the report's eventual use. The monthly report on discharge license parameters is discussed further on in the report.

The monthly operations report is used for operational review and planning and, therefore, requires considerably more information than effluent quality. This report will include the final effluent data but will also include all laboratory data related to plant processes. In addition to laboratory data, the report could include return sludge rates, mass of sludge removed or accumulated, loading parameters, chemicals added, temperatures, and critical operational log entries (for example, equipment failures or mode changes).

BILLING REPORTS. Many communities are faced with significant industrial or commercial discharges to their sewer systems. Therefore, billing reports are often generated from the plant data as part of an industrial pretreatment program. In some cases, the only analytical indication that industrial or commercial wastes are significant is the change in wastewater color.

The only way to determine the industrial sector's appropriate contribution to operational costs is to have accurate and representative sampling of the individual waste streams. The determination of representative samples should follow the guidelines in this manual, but special attention also should be given to sewer-use ordinances and permit requirements that may dictate special conditions. In some cases, the type and frequency of monitoring may be established by mutual agreement between the community and the industry as long as the sampling has been shown to be representative.

Other special considerations of industrial monitoring include the use of bound notebooks for logging and analyzing data. If other plant or city personnel are used for sample collection, consideration should be given to chain-of-custody procedures.

The important consideration is the sensitivity of the particular data. If the daily loadings from a particular industry are fairly steady, there should not be much controversy over the data. If, however, the industry is prone to wide

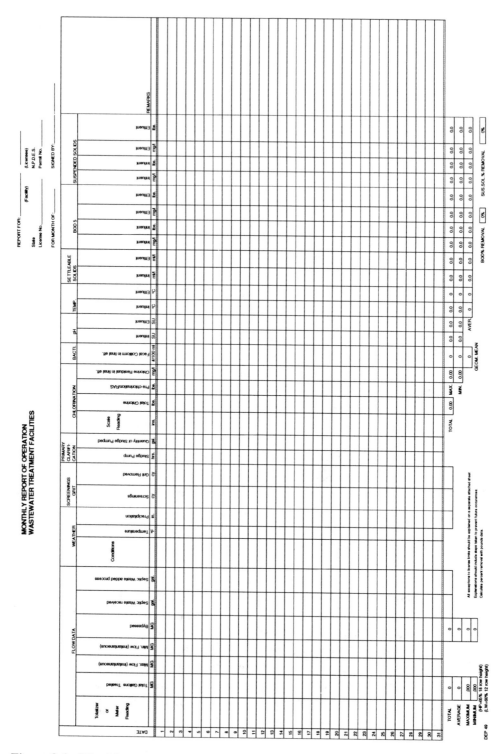

Figure 8.6   Monthly report.

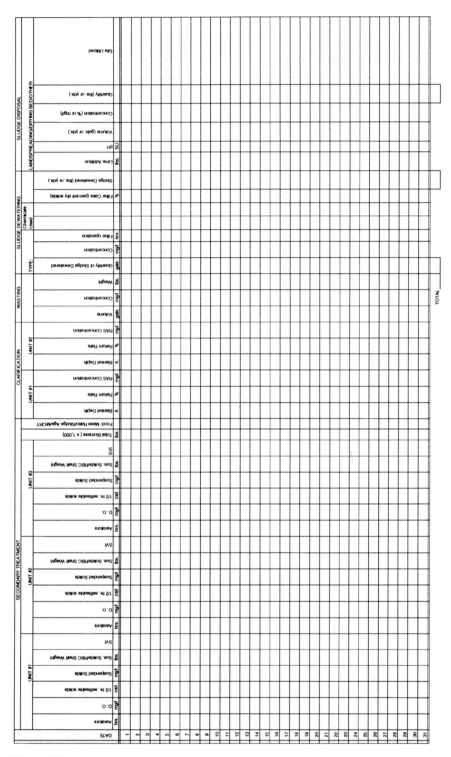

Figure 8.7 Monthly summary.

and varying ranges of loadings that would make it susceptible to punitive billing actions, then the data should be handled with extra care.

Reporting considerations will also vary according to the particular industry, the nature of the wastewater, and any other special billing requirements. Typically, however, the billing report should include the number of individual data points (frequency of sampling) and average, maximum, and minimum values (Figure 8.8). The report should be signed by both the laboratory manager and the plant manager and then forwarded to the appropriate regulatory office, such as the city clerk's office or a public works department. Although monthly reports typically are appropriate for billing, Figure 8.9 is an annual report used to show seasonal trends of industries A, B, and C. This table highlights the frequency of a particular event.

Often, only the monthly total industrial load is needed for billing purposes. However, in special cases, individual days or peaks will be important. Reporting forms should reflect the data that are significant to the billing procedures. The monthly billing report assumes that the cost of treatment is a function of the treatment cost per unit mass of BOD plus the treatment cost per unit mass of suspended solids. In special cases, other considerations may have to be factored into the cost of treating industrial wastes. When these factors exist, they should be included in the report.

REGULATORY REPORTS. One significant aspect of water pollution control legislation in the U.S. is the National Pollutant Discharge Elimination System (NPDES) permit program. In general, all point source dischargers in the U.S. must obtain a permit authorizing and regulating the discharge contents. Most states have assumed primary enforcement responsibility for the NPDES program. States participating in the NPDES program must ensure that issued permits comply with national effluent limitations, new source per-

Billing Report Form

Parameter	Concentration			Quantities			Frequency of	Analysis
	Max.	Min.	Avg.	Max.	Min.	Avg.		
Flow								
BOD								
TSS								
pH								

Monthly Cost = $ _____ lb/BOD + $ _____ /lb TSS = _____
Plant Manager: _____   Lab Manager: _____
Date: _____   Date: _____

Figure 8.8   Typical billing report form (lb × 0.453 6 = kg).

Industrial Summary Report

Industry		A	B	C	Total Industrial	Total Plant
FLOW, gal	Jan. Feb. Mar. Apr. May June July Aug. Sept. Oct. Nov. Dec. Avg.					
BOD, lb	Jan. Feb. Mar. Apr. May June July Aug. Sept. Oct. Nov. Dec. Avg.					
TSS, lb	Jan. Feb. Mar. Apr. May June July Aug. Sept. Oct. Nov. Dec. Avg.					

Figure 8.9 Typical industrial summary report (gal $\times$ [3.785 $\times$ 10^{-3}] = m^3; lb $\times$ 0.453 6 = kg).

formance, and toxic standards. In addition, these states may impose standards and requirements more stringent than those of the federal government.

The NPDES permit contains three major elements: compliance schedules, effluent limits, and self-monitoring requirements.

The compliance schedule empowers the regulatory agency to require and monitor specific actions of the discharger. The compliance schedule may contain dates for achieving certain levels of progress on activities that will improve treatment efficiencies. Those activities may include engineering reports, final construction plans, construction status reports, and operation commencement schedules.

The second major element of the NPDES permit is its use as a vehicle for the regulatory agency to communicate effluent requirements to a discharger. Minimum effluent requirements typically limit the quantities of discharged suspended solids and oxygen-demanding materials. Other requirements may be imposed by regulatory agencies to protect the quality of receiving waters.

The third element of the NPDES permit is the required monitoring and reporting of the quality of effluent discharged. The self-monitoring requirements may vary from discharger to discharger, but they all address the following:

- Parameters to be monitored,
- Sample types (grab or composite),
- Sampling frequency,
- Flow measurement frequency,
- Analytical methods, and
- Frequency of reporting to the regulatory agency.

One important fact to remember is that the NPDES permit is a legal document and is binding on the permit holder. The original permit should be kept in a safe and secure place, and a copy should be kept at the plant for easy reference. The permit should be consulted often to determine whether all conditions are being met. Stiff penalties may be imposed for failure to comply with the permit. If questions arise, they should be addressed immediately to the local state regulatory agency or nearest U.S. EPA regional office. Compliance reports should be submitted regularly to the state or U.S. EPA agency that has jurisdiction in the discharger's area.

Compliance reports vary from state to state. Figure 8.10 is an example of a U.S. EPA discharge monitoring report. The following are general comments for filling out a compliance report; the local regulatory agency should be consulted for specific requirements.

- The time period covered by the report should be entered.
- If no discharge occurs during the report period, a "no discharge" status should be reported.
- If analysis results for effluent parameters or tests exceed those specified in the permit, this information also must be reported.
- Some permits will call for 30-day averages, 7-day averages, and daily maximum limits. Exceeding any of these limits will constitute a violation of the permit.
- A geometric mean should be used for averaging bacterial counts.
- Percentage removal requirements must be calculated without influence of recycle streams:

$$\% \text{ Removal} = \frac{\text{Influent concentration} - \text{Effluent concentration}}{\text{Influent concentration}} \times 10$$

- Methods of analysis and sampling types and locations will be specified in the permit. In some cases, shortened time intervals will be allowed for composite sampling.

WRD 2A-82

Winter Limitations (November 1 through April 30)

STATE OF WEST VIRGINIA
NATIONAL POLLUTANT DISCHARGE ELIMINATION SYSTEM
DISCHARGE MONITORING REPORT

COMMERCIAL LABORATORY NAME _____
COMMERCIAL LABORATORY ADDRESS _____

FACILITY NAME _____
LOCATION OF FACILITY _____
PERMIT NUMBER _____   OUTLET NO. _____ 19
WASTELOAD FOR MONTH OF _____

INDIVIDUAL PERFORMING ANALYSES _____

Parameter		Quantity					Other Units					Measurement Frequency	Sample Type
		Minimum	Avg. Monthly	Max. Daily	Units	N.E.	Minimum	Avg. Monthly	Max. Daily	Units	N.E.		
Flow, in Conduit or thru trmt. plant 50050	Reported	.....	....	....	....	..	.....	....	....	MGD	..		
	Permit Limitation	.....	....	....	....	..	N/A	3.5	N/A		..	Continuous	Measured
BOD, 5 day (20 Deg. C) 00310	Reported	.....			lbs/day	..	.....			mg/l	..		
	Permit Limitation	N/A	875.7	1,751	lbs/day	..	N/A	30.0	60.0		..	Twice/Week	8 Hour Composite
Solids, Total Suspended 00530	Reported	.....	....	....	lbs/day	..	.....			mg/l	..		
	Permit Limitation	N/A	875.7	1,751	lbs/day	..	N/A	30.0	60.0		..	Twice/Week	8 Hour Composite
Nitrogen, Total Kjeldahl 00625	Reported	.....	....	....	lbs/day	..	.....	....	....	mg/l	..		
	Permit Limitation	N/A	525.4	1,051	lbs/day	..	N/A	18.0	36.0		..	Twice/Week	8 Hour Composite
pH 00400	Reported	.....	....	....	....	..	.....	....	....	Std. Units	..		
	Permit Limitation	.....	....	....	....	..	6.0	N/A	9.0		..	Daily	Grab
Coliform, Fecal General 74055	Reported	MF	- - -	MPN	....	..	.....	....	....	Count per 100 ml	..		
	Permit Limitation	Circle	Method	Used	....	..	N/A	200	400		..	Twice/Week	Grab
Oxygen, Dissolved 00300	Reported	.....	....	....	....	..	.....	....	....	mg/l	..		
	Permit Limitation	.....	....	....	....	..	6.0	....	....		..	Twice/Week	Grab

Name of Principal Executive Officer _____

Title of Officer _____

I certify under penalty of law that this document and all attachments were prepared under my directions or supervision in accordance with a system designed to assure that qualified personnel properly gather and evaluate the information submitted. Based on my inquiry of the person or persons who manage the system, or those persons directly responsible for gathering the information, the information submitted is, to the best of my knowledge and belief true, accurate and complete. I am aware that there are significant penalties for submitting false information including the possibility of fine and imprisonment for knowing violations.

Date Completed _____

Signature of Principal Executive Officer or Authorized Agent _____

Figure 8.10   West Virginia NPDES discharge monitoring report.

STATE OF WEST VIRGINIA
NATIONAL POLLUTANT DISCHARGE ELIMINATION SYSTEM
DISCHARGE MONITORING REPORT

Winter Limitations (November 1 through April 30)

FACILITY NAME _____
LOCATION OF FACILITY _____
PERMIT NUMBER _____ OUTLET NO. _____ 19
WASTELOAD FOR MONTH OF _____

COMMERCIAL LABORATORY NAME _____
COMMERCIAL LABORATORY ADDRESS _____
INDIVIDUAL PERFORMING ANALYSES _____

Parameter		Quantity					Other Units					Measurement Frequency	Sample Type
		Minimum	Avg. Monthly	Max. Daily	Units	N.E.	Minimum	Avg. Monthly	Max. Daily	Units	N.E.		
Chlorine, Total Residual 50060	Reported	.....	....	....	...	...	......	......			...		
	Permit Limitation	N/A	N/A	N/A	...	...	N/A	N/A	0.010	mg/l	...	Twice/Week	Grab
Cadmium, Total 01027	Reported	.....	....	....	...	...	......	......			...		
	Permit Limitation	N/A	N/A	N/A	...	...	N/A	N/A	0.0011	mg/l	...	1/Month	8 Hour Composite
Chromium, Hexavalent 01032	Reported	.....	....	....	...	...	......	......			...		
	Permit Limitation	N/A	N/A	N/A	...	...	N/A	N/A	0.0100	mg/l	...	1/Month	8 Hour Composite
Copper, Total 01042	Reported	.....	....	....	...	...	......	......			...		
	Permit Limitation	N/A	N/A	N/A	...	...	N/A	N/A	0.0110	mg/l	...	1/Month	8 Hour Composite
Lead, Total 01051	Reported	.....	....	....	...	...	......	......			...		
	Permit Limitation	N/A	N/A	N/A	...	...	N/A	N/A	0.0032	mg/l	...	1/Month	8 Hour Composite
Nickel, Total 01067	Reported	.....	....	....	...	...	......	......			...		
	Permit Limitation	N/A	N/A	N/A	...	...	N/A	N/A	0.1577	mg/l	...	1/Month	8 Hour Composite
Silver, Total 01077	Reported	.....	....	....	...	...	......	......			...		
	Permit Limitation	N/A	N/A	N/A	...	...	N/A	N/A	0.0040	mg/l	...	1/Month	8 Hour Composite

Name of Principal Executive Officer _____

Title of Officer _____

I certify under penalty of law that this document and all attachments were prepared under my directions or supervision in accordance with a system designed to assure that qualified personnel properly gather and evaluate the information submitted. Based on my inquiry of the person or persons who manage the system, or those persons directly responsible for gathering the information, the information submitted is, to the best of my knowledge and belief true, accurate and complete. I am aware that there are significant penalties for submitting false information including the possibility of fine and imprisonment for knowing violations.

Date Completed _____

Signature of Principal Executive Officer or Authorized Agent _____

Figure 8.10   West Virginia NPDES discharge monitoring report (continued).

WRD 2A-82

Winter Limitations (November 1 through April 30)

STATE OF WEST VIRGINIA
NATIONAL POLLUTANT DISCHARGE ELIMINATION SYSTEM
DISCHARGE MONITORING REPORT

COMMERCIAL LABORATORY NAME _____
COMMERCIAL LABORATORY ADDRESS _____

INDIVIDUAL PERFORMING ANALYSES _____

FACILITY NAME _____
LOCATION OF FACILITY _____
PERMIT NUMBER _____   OUTLET NO. _____ 19 ____
WASTELOAD FOR MONTH OF _____

Parameter		Quantity					Other Units					Measurement Frequency	Sample Type
		Minimum	Avg. Monthly	Max. Daily	Units	N.E.	Minimum	Avg. Monthly	Max. Daily	Units	N.E.		
Zinc, Total 01092	Reported	.....	....	......	...	..	.....	....	....		..		
	Permit Limitation	N/A	N/A	N/A	...	..	N/A	N/A	0.0500	mg/l	..	1/Month	8 Hour Composite
Mercury, Total 71900	Reported	.....	....	......	...	..	.....	....	....		..		
	Permit Limitation	N/A	N/A	N/A	...	..	N/A	N/A	0.0002	mg/l	..	1/Month	8 Hour Composite
											..		

Name of Principal Executive Officer

Title of Officer

I certify under penalty of law that this document and all attachments were prepared under my directions or supervision in accordance with a system designed to assure that qualified personnel properly gather and evaluate the information submitted. Based on my inquiry of the person or persons who manage the system, or those persons directly responsible for gathering the information, the information submitted is, to the best of my knowledge and belief true, accurate and complete. I am aware that there are significant penalties for submitting false information including the possibility of fine and imprisonment for knowing violations.

Date Completed

Signature of Principal Executive Officer or Authorized Agent

Figure 8.10   West Virginia NPDES discharge monitoring report (continued).

- Acceptable quality assurance procedures should be used and documented to support reliability of all data reported.
- The reports must be signed by an authorized signing official; if more than one page is submitted, then each page should be signed.
- pH values should not be averaged.
- The permit prohibits discharging floating oil or grease or any toxic materials. Any discharges of these materials should be reported immediately.
- The permit prohibits bypassing of treatment facilities except to prevent loss of life or severe property damage. Any bypassing should be reported immediately.

The law requires that regulatory monitoring data be kept for a specific time period, typically 3 years (though this requirement varies from state to state). Self-monitoring records include all bench sheets, daily summaries and workbooks, and all quality assurance data generated on the analyses. All monthly reports must be maintained for the designated period.

SPECIAL REPORTS. Special reports will be required from time to time by different groups of people, and a specific format may not be provided. In such cases, the organization of the laboratory program will pay off. The bench sheets, daily summaries, and monthly reports will have been filed in the workbook for ready use. If a computer system is available, a simple program might be used to retrieve the data.

First, before any data are retrieved, the request should be considered carefully. What data are needed to complete the report? Are these particular data available? Are any special tests needed to complete the report? After these questions have been answered, the report can be outlined. Following are four different outlines for four special reports that might be encountered.

Process Reports. The plant manager or regulatory agencies may require information to evaluate a particular event. An example of how to evaluate a request follows:

- What is the problem?
  - — Evaluate the microscopic changes that have occurred during the past month in the activated-sludge system.
- What parameters are needed?
  - — Organic loading conditions,
  - — Nutrient loading conditions,
  - — Solids inventory changes,
  - — Microscopic analysis reports, and
  - — Process control trends.

- What data are needed?
  - — Modes of identifying filamentous organisms, and
  - — Information on special operational situations, such as problems with dissolved oxygen.
- In what way must the data be presented?
  - — Evaluated data for any relationship to the parameter identified.
- Are existing data sufficient, or are more data needed?
  - — Collect more data if needed.
- Solution:
  - — Present the analysis of the problem and support conclusions with graphs or statistical analyses of data.

Cost Analysis Reports. Cost analysis reports may be requested by the plant manager or public works department for a variety of subjects such as chemical costs, energy costs per unit process, and solids-handling costs. An example outline of such a report follows:

- What is the problem?
  - — Evaluate the cost of incinerating different mixtures of sludge.
- What parameters are needed?
  - — Volatile solids of sludge streams,
  - — Percent moisture of sludge streams,
  - — Quantity of sludge streams, and
  - — Ratio of sludge streams to be incinerated.
- What data are needed?
  - — Kilojoules per kilogram (British thermal units per pound) volatile solids, and
  - — Cost per kilojoule (British thermal units) of fuel.
- In what way must the data be presented?
  - — Graph the kilojoules (British thermal units) of fuel needed for the different mixtures of sludge and concentrations of sludge.
- Are existing data sufficient, or are more data needed?
  - — Collect more data if needed.
- Solution:
  - — Identify the most probable range of operating conditions and the cost associated with this range.

Noncompliance Reports. Occasionally, something will happen in the plant to cause a violation of the permit. When a violation occurs, a report will be required. An example outline would be the following:

- What is the problem?
  - — Fecal coliforms were found to be high for 5 consecutive days.

- What parameters are needed?
  — Effluent quality, and
  — Chlorine demand test.
- What data are needed?
  — Information on special operating conditions such as cleaning of contact tanks; and
  — Description of unusual laboratory procedures or conditions, such as bad media.
- In what way must the data be presented?
  — Split samples with regulatory agency or commercial lab, and
  — Evaluate effects of changing chlorine feed rates or lab procedures.
- Are existing data sufficient, or are more data needed?
  — Collect more data if needed.
- Solution:
  — Discuss the degree of the violations, the measures taken to determine the cause of violations, and the corrective action taken to eliminate the violations.

Historical Reports. During plant expansion or major modifications, special reports may be required to project future events. An example outline might be as follows:

- What is the problem?
  — What is a realistic flow projection for the next 20 years?
- What parameters are needed?
  — Past flow data;
  — Population served;
  — Industrial contributors; and
  — Future projects, major industrial plants, or city growth plans.
- What data are needed?
  — Rainfall data, and
  — Any information on unusual conditions such as grouting collection lines.
- In what way must the data be presented?
  — Compare the plant flow records with the parameters evaluated.
- Are existing data sufficient, or are more data needed?
  — Collect more data if needed.
- Solution:
  — Determine realistic flow per person treated in the system.

DATA REPORTING MANAGEMENT SYSTEMS

Quality data reporting is a management system from the bench sheets to the final report. Outlined in this chapter is a manual data reporting system, which is adequate for many facilities, although a computerized system adds another dimension of management to data handling and reporting.

A laboratory information management system (LIMS) is often used by larger facilities but can also be modified for smaller plants. When using a computerized system, bench sheets and workbooks are still required according to U.S. EPA. The advantages of computerized data handling and reporting include rapid transmission of data, immediate data analysis through graphing, fewer mathematical and transcription errors, projecting process trends, report generation, and the ability to link reports (that is, monthly summaries and discharge monitoring reports). Laboratory information management systems offer advantages to larger laboratories with gas chromatograph and mass spectrophotometer equipment. However, any system should be set up so that information can easily be entered from bench sheets to the computer. One great advantage of the LIMS can be computer-assisted process control for developing mass balances, process loadings, and other reports. Statistical analyses also can be run on plant data and used to develop control strategies. Figure 8.11 is an example of an LIMS form.

Computer data handling has its disadvantages, however. One major disadvantage is the ease with which data can be changed or lost. Nevertheless, there are methods to reduce this risk. For example, storing data on backup disks, filing bench sheets and workbooks as required by regulatory agencies and the owner, establishing audit trails, and creating password-triggered access that is based on authority levels (that is, read-only privileges, data entry, but not validation privileges). Thus, with proper care, the danger can be minimized, and the computer becomes a powerful instrument that not only saves time but also allows for easier preparation of reports and statistical analyses.

All data and reports generated at wastewater treatment plants become available to the public and third-party interest groups. Therefore, it is important to be thorough and completely accurate for both regulatory and public relations concerns.

USA WWTF GENERAL LABORATORY		
WATER SECTION	DAILY SAMPLE LOG	
Sample Date _____	Shift Leader _____	Analysis Date _____

Sample Type: BRELUTE

Sample ID	Rec'd (Y/N)	LIMS NUMBER	Sample ID	Rec'd (Y/N)	LIMS NUMBER
MB			JB1		
GCEC			OFB		
MV#1			96SC		
EC					

Sample ID	Rec'd (Y/N)	LIMS NUMBER	Sample ID	Rec'd (Y/N)	LIMS NUMBER
INFL SLDG TK 1			GST EFFL COMP		
INFL SLDG TK 2			DAF EFFL 1		
EFFL RETN TK 1			DAF EFFL 2		
PELLETIZER CENTRATE			DAF EFFL 3		
			DAF EFFL 4		

Sample ID	Rec'd (Y/N)	LIMS NUMBER	Sample ID	Rec'd (Y/N)	LIMS NUMBER
MIXED LIQ 1			MIXED LIQ 6		
MIXED LIQ 2			MIXED LIQ 7		
MIXED LIQ 3			MIXED LIQ 8		
MIXED LIQ 4			MIXED LIQ 9		
MIXED LIQ 5			MIXED LIQ 10		

Figure 8.11   Laboratory information management system log-in sheet, water and sludge.

	USA WWTF GENERAL LABORATORY				
WATER SECTION		DAILY SAMPLE LOG			
Sample Date _____		Shift Leader _____		Analysis Date _____	

Sample ID	Rec'd (Y/N)	LIMS NUMBER	Sample ID	Rec'd (Y/N)	LIMS NUMBER
GST INFL COMP			CENTRATE 1		
DAF INFL COMP			CENTRATE 2		
			BELTPRESS- FILTRATE		

Sample ID	Rec'd (Y/N)	LIMS NUMBER	Sample ID	Rec'd (Y/N)	LIMS NUMBER
SEPTIC 1			SEPTIC 6		
SEPTIC 2			SEPTIC 7		
SEPTIC 3			SEPTIC 8		
SEPTIC 4			SEPTIC 9		
SEPTIC 5			SEPTIC 10		

Sample ID	Rec'd (Y/N)	LIMS NUMBER	Sample ID	Rec'd (Y/N)	LIMS NUMBER
DIGESTER GAS C1			DIGESTER GAS C5		
DIGESTER GAS C2			DIGESTER GAS C6		
DIGESTER GAS C3			DIGESTER GAS E7		
DIGESTER GAS C4			DIGESTER GAS E8		

Sample ID	Rec'd (Y/N)	LIMS NUMBER	Sample ID	Rec'd (Y/N)	LIMS NUMBER

Figure 8.11   Laboratory information management system log-in sheet, water and sludge (continued).

```
┌─────────────────────────────────────────────────────────────────────────┐
│                    USA WWTF GENERAL LABORATORY                            │
│  WATER SECTION                  DAILY SAMPLE LOG                          │
│  Sample Date _____    Shift Leader _____   Analysis Date _____│
└─────────────────────────────────────────────────────────────────────────┘
```

Sample Type: THICKENERS

Sample ID	Rec'd (Y/N)	LIMS NUMBER	Sample ID	Rec'd (Y/N)	LIMS NUMBER
DAF SLDG 1			DAF POLYMER 1		
DAF SLDG 2			DAF POLYMER 2		
DAF SLDG 3			CENT POLYMER 1		
GST SLDG COMP					

Sample Type: SLUDGE-1

Sample ID	Rec'd (Y/N)	LIMS NUMBER	Sample ID	Rec'd (Y/N)	LIMS NUMBER
PRIM SLDG 1-7			GST SCUM		
PRIM SLDG 8-11			CENT NO 1 FEED		
ACT SLDG 1-2			CENT NO 2 FEED		
ACT SLDG 3-4			BELT PRESS FEED		
ACT SLDG 5-10			CENT NO 2 CAKE		
GRIT CHAMBER			DIGESTED SLDG		
CYCLONE DEGRIT					

Figure 8.11 **Laboratory information management system log-in sheet, water and sludge (continued).**

SLUDGE SECTION　　　　　　　　DAILY SAMPLE LOG

Sample Date _____　　Shift Leader _____　　Analysis Date _____

Sample Type: DIGESTER

Sample ID	Rec'd (Y/N)	LIMS NUMBER	Sample ID	Rec'd (Y/N)	LIMS NUMBER
DIG C1			DIG C5		
DIG C2			DIG C6		
DIG C3			DIG E7		
DIG C4			DIG E8		

Sample Type: COMPOSITE

Sample ID	Rec'd (Y/N)	LIMS NUMBER	Sample ID	Rec'd (Y/N)	LIMS NUMBER
CENT NO 1 CAKE			7 DAY CENT NO 1 CAKE		
BELT PRESS SLDG					

Sample Type: BREFFLUENT (COLIFORM SAMPLE)

Sample ID	Rec'd (Y/N)	LIMS NUMBER	Sample ID	Rec'd (Y/N)	LIMS NUMBER
OFB					

Sample Type: BCS (COMPOSTING FACILITY)

Sample ID	Rec'd (Y/N)	LIMS NUMBER	Sample ID	Rec'd (Y/N)	LIMS NUMBER
BCS			BCS		
BCS			BCS		

Figure 8.11 **Laboratory information management system log-in sheet, water and sludge (continued).**

Chapter 9
Safety

COMMITMENT TO SAFETY

COMPANY COMMITMENT. There must be a commitment to safety in any organization. Management must be willing to spend the necessary time and money to achieve stated safety goals. To be effective, the safety program must be the number one priority of both management and employees. Safety in the laboratory and in the field is a continuing commitment that requires effort and vigilance. A good program should consist of an organizational safety policy statement reinforced by detailed and enforceable rules.

Policy Statement. The policy statement should be well planned, discussed, and written. It must include management's position on safety in the laboratory and in the field, as well as a brief description of the rules and regulations necessary to achieve this position. This description should include a plan for regularly scheduled safety meetings to evaluate effectiveness. A safety manager/director should be appointed with the responsibility of implementing the safety program to meet the stated goals. The policy statement does not have to be long but must be written in definite terms so that there is no doubt in a laboratory or field worker's mind as to the importance of following the policy in everyday work activities. This policy statement should be the first page of any laboratory hazard communication program or safety standard operating procedure.

Safety Rules. Safety rules must be clearly written and enforceable. In many cases, safety rules are really common sense guidelines. However, they must also be complete to satisfy both the policy statement and recent governmental regulations. Two recent governmental regulations that have affected laboratory safety programs are 29 CFR Part 1910, *Occupational Exposure to Hazardous Chemicals in Laboratories, Final Rule*, and 29 CFR Part 1910.1030, *Occupational Exposure to Bloodborne Pathogens*. Both rules were written by the Occupational Safety and Health Administration (OSHA) and require the laboratory safety director to evaluate the effect of each regulatory requirement on safety standard operating procedure. Typically, changes to standard safety rules are required to comply with the published standards.

The laboratory and field safety rules should consider the use of safety equipment, indoctrination of new employees, hazard communication, housekeeping, normal work procedures, new work procedures, the reporting of incidents and accidents, and personal responsibility. Occasionally, there may be valid extenuating circumstances that require the reevaluation of a safety rule. However, when there are no extenuating circumstances, the rules must be consistently enforced.

Training and Testing. An often overlooked area in safety planning is new testing and alterations to existing training. Over time, both management and employees become familiar with the existing safety program, and the tendency toward complacency exists. For this reason, it is important to apply the laboratory and field safety rules *before* commencing any new testing. Also, alterations to testing procedures must be approved by the laboratory safety director. Many alterations to testing, particularly in production laboratories, are made to save time and/or materials. It is important that these changes be impartially evaluated to ensure that safety has not been compromised in the process. One laboratory recently found this out when the use of a water bath for an acid digestion of water samples was replaced with direct heating on a hot plate. The glass bottles used for the heating were not designed to withstand the heat and pressure that the use of the hot plate created, and the bottles exploded in the hood. While the hood contained most of the damage and no one was injured, the potential for property damage and personal injury was created. Policies requiring the use of plastic bottles and a water bath for acid digestion were immediately instituted.

Safety Equipment. Laboratory and field safety equipment must be available and in good working order, and technicians should be trained in the proper use of all items during indoctrination. Safety equipment can be separated into two areas: routine and emergency. Routine safety equipment includes a laboratory coat or apron or linear polyethylene fiber suit or coveralls, safety glasses with side shields, splash goggles, proper protective gloves, face shields for use with concentrated acids and bases, respirators, and protective bottle carriers. In addition, approved department of transportation containers should be used for field work, and fume hoods and proper storage cabinets should be used in the laboratory. The use of these items is fairly routine but must be stressed from the beginning. In fact, incorporating everyday safety items should become second nature for all laboratory and field workers.

Emergency safety equipment includes showers, eyewash stations, fire extinguishers, fire blankets, emergency exits/alarms, and first-aid kits. The use of these items may not be routine, and proper training must begin during indoctrination. The use of safety showers and eyewashes should be demonstrated, and items in the first-aid kit should be explained. For laboratory work, emergency exits and alarms must be located and all functions properly understood.

Emergency escape routes and alarm systems must be discussed and documented before conducting any field work. The basics of fire extinguisher use should be taught. It is helpful and educational to have a volunteer fire company demonstrate how to put out a small fire with a fire extinguisher. Because the use of these items will not become second nature, periodic reinforcement at safety meetings will be required.

The maintenance of each piece of safety equipment is important. Individual safety equipment must be maintained by each laboratory and field worker. Laboratory-owned equipment must receive periodic maintenance performed under the supervision of the safety director. Such maintenance includes the regular measurement of fume hood velocities, cleaning and testing of safety showers and eyewashes, and annual inspection of fire extinguishers. The location of all equipment must be pointed out to all new laboratory technicians.

PERSONAL RESPONSIBILITY. Personal responsibility among laboratory and field technicians is an important aspect of a commitment to safety. It is each individual's responsibility to follow the safety program and be aware of his or her work environment. Individuals should follow procedures without modifications, and any contemplated changes must be discussed with the designated safety personnel. It is important to remember that modifications that appear to be trivial may potentially be lethal.

Participation should be encouraged by management, and good ideas should be rewarded. The technician performing a test repetitively typically will develop insight on how to improve the safety procedures that govern the workplace. The ideas generated by this person can be promoted through a suggestion box or at scheduled safety meetings. The promotion of these ideas from laboratory and field technicians through management will aid in creating a safe working environment and improve overall morale.

*I*NDOCTRINATION

All new trainees must report to the safety director before commencing any laboratory or field work. At this time, the safety director should explain the laboratory and field safety program and have the new employee read the laboratory standard operating procedures for safety. After any preliminary questions about the safety program are answered, a tour of the laboratory should be given. Special emphasis should be placed on the location of all safety equipment in the laboratory and the locations of emergency exits and manual alarms. The safety director should be confident that the basics of safe operations in the laboratory are understood. More preliminary training may be required depending on the level of formal education and/or previous laboratory experience. A written test can be administered to validate the initial training, if desired. Once the preliminary safety training is completed, personal safety equipment should be distributed.

The indoctrination continues by designating an established employee as a "buddy." Every new employee needs someone to explain other relevant standard operating procedures. When the training progresses to the point where actual testing procedures will be performed, a review of the specific safety procedures must be undertaken. The standard operating procedure for the testing being performed should also include any specific safety precautions necessary, as well as the possible consequences of negligence.

Another key element of the indoctrination period involves the initial review by the trainee of the material safety data sheets (MSDS) for the chemicals used in the procedure. Evaluate the toxicity of all chemicals before handling or testing. Any questions regarding the MSDS or its interpretation should be answered by the safety director.

These initial safety training steps promote the importance of safety in the laboratory and field and provide new trainees with the proper information necessary to avoid injury in the workplace.

*H*OUSEKEEPING

Housekeeping in the laboratory encompasses the orderly arrangement of all reagents, equipment, and supplies. Good housekeeping practices promote safety and develop efficient habits that augment effective performance. Each person is responsible for keeping his or her own work area clean. Good housekeeping in the laboratory is critical to safe operations.

Common problems related to housekeeping include tripping hazards, piles of paper and office supplies on top of cabinets, obstructed exits, glassware piled on carts, and wet floors.

The following guidelines apply to good housekeeping:

- Provide a definite storage place for each item.
- Return all instruments, equipment, or reagents to the proper location immediately after use.
- Keep aisles, walkways, halls, and exits free of obstructions and tripping hazards.
- Clean up, immediately, any liquids or solids spilled on the floor or work bench. Mark any spill with a "wet floor" sign.
- Keep sinks and hoods clean and clear. Do not use hoods as long-term storage areas.
- Wash glassware often to prevent unnecessary accumulations.
- Deposit broken glassware in designated containers.
- Know the location of fire extinguishers, spill kits, first-aid kits, fire blankets, emergency showers, eyewash stations, and emergency exits. Keep these areas clear and unobstructed at all times.
- Close all cabinet drawers immediately after use.

*L*ABORATORY WORK PROCEDURES

FOOD AND BEVERAGES. Food and beverages should not be allowed in the laboratory area under any circumstances. Accidental or airborne contamination is possible, with potentially harmful consequences. Likewise, where eating and drinking is allowed, laboratory items (reagents, coats, and gloves, for example) should not be present. Laboratory workers should be reminded to thoroughly wash their hands before breaks and lunches.

GLASSWARE. Accidents resulting from the use of broken or improper glassware are probably the most commonly encountered hazards in the laboratory. Technicians sometimes use broken glassware because of budgetary constraints and an unwillingness to order replacements. This will eventually result in failure of the glassware during operation or lacerations from handling the glassware. While the accidents are often minor, serious puncture wounds can result and create the potential for chemical exposure through the wound. Also, serious injuries from explosions resulting from failed or improper glassware are possible.

The easiest way to prevent these injuries is to remove broken glassware from circulation. This function should be performed by the personnel using the glassware. It is also helpful to have the personnel cleaning the glassware remove broken items. If the glassware is repairable, it should be sent to a company qualified to make the repair. A log detailing all broken glassware should be kept so that replacements can be ordered in a timely fashion. It may also be possible and economically desirable to replace pieces traditionally made of glass with materials that are less fragile and of similar chemical resistance.

The glassware used for an analysis must be of the right construction. If an operation is being performed at a high temperature with concentrated solutions, the glassware must meet both the temperature and chemical-resistance requirements. Ignition test tubes and heavy-walled beakers are examples of reinforced glassware for chemical testing.

CHEMICALS. Proper chemical use and storage in the laboratory is an important part of any chemical hygiene program. Much of the information needed to safely handle chemicals is provided in the MSDS. However, for the MSDS to be useful, it must be understood. The safety director can assist in interpreting the information and making informed decisions.

Many chemicals presently available for purchase come with labels that provide a synopsis of the information found in the MSDS. These labels are useful reminders to a technician who has used and is familiar with the chemical. While it is not a replacement for reading the MSDS, the label can aid the technician by providing quick, ready reference to storage requirements, handling procedures, and hazard ratings. These storage and hazard guidelines differ between chemical manufacturers, so attention to detail is necessary.

The basic ingredients of a laboratory chemical hygiene program include

- Education and training of employees in the requirements of the laboratory's hazard communication program,
- Use of MSDS and other literature to determine associated hazards and necessary safety precautions when chemicals are used,
- Use of MSDS and other literature to determine proper storage of chemicals when not in use, and
- Internal labeling program to identify the associated hazards and proper storage of solutions or aliquots of the chemicals.

Proper storage areas should be available for acids and bases, flammable materials, and dry chemicals. Commercial cabinets are available that meet regulatory specifications for the storage of flammable liquids and acids or bases. Dry reagents should be stored appropriately, segregating general storage chemicals from oxidizers and reducers. Poisons should be kept in an area of limited access, preferably under lock and key.

Flammable liquids must be stored in properly grounded storage cabinets or drums. When they are in use, the quantity of the liquids should be limited.

These liquids should be transported in plastic carriers to prevent accidental breakage. All use of flammable chemicals should be done in areas that have exhaust hoods or are well ventilated. These chemicals should never be heated over an open flame or transferred in areas where hotplates are in operation.

Proper general chemical use requires that the technician consider the toxic potential of handling and using the chemical. Material safety data sheets for every chemical must be kept in a file or binder in alphabetical order. When an updated MSDS is received from a manufacturer, it must be used in place of the existing one.

Several other sources of literature for determining chemical hazards are available at local public or university libraries. Publications found to be especially useful should be purchased and placed in the laboratory library. Of special interest are reactivity data and information about the carcinogenic effects of chemicals. Other considerations when handling chemicals are as follows:

- Be sure to use the proper method of confinement when working with a chemical. If breaking the container would create a hazard, use proper carriers or spill trays.
- Limit the quantity of chemicals in use at one time. This will help avoid confusion and misidentification.
- Use fume hoods whenever possible to avoid toxic vapors.
- Always use the appropriate personal safety equipment.
- Dispose of waste chemicals, soiled gloves, pipets, and other items appropriately. When in doubt, ask the safety supervisor.
- Properly label any reagents created.
- Always add concentrated chemicals to water.
- Clean up all spills immediately.

FUME HOODS. Fume hoods should be used whenever possible, but particularly when the potential exists for the release of toxic chemicals through volatilization. In addition to controlling harmful vapors, fume hoods can be used for emergency containment in the event something goes wrong with a laboratory operation. By keeping the sash down when a technician is not physically working in a fume hood, flying objects from shattered glassware, splashed chemicals, and small fires can be contained. Air velocity through the fume hood is also increased when the sash is down, thereby ensuring a negative pressure in the area where the chemicals are used.

The air velocity should be measured regularly, including every time any adjustment or mechanical work is performed. Instruments are now available that can continuously monitor air velocity in a fume hood and notify personnel by sound alarm if velocities fall beneath set limits. If a hood is failing, mechanical maintenance of the blower or motor may be required.

One precautionary note is important when using fume hoods. Often, the laboratory door is closed and the whole building may be under negative pressure. This reduces velocities within the fume hood and, if the low velocities cannot be mechanically improved, the installation of a makeup air system may be required.

CHEMICAL SPILLS. When a chemical spill occurs in the laboratory, certain actions are necessary to ensure that the spill is cleaned and disposed of properly. Laboratory and field sampling workers cannot afford to be nonchalant about any spill, even water. *All spills must be cleaned up immediately.* Also, remember that the material used to clean up the spill inherits the properties of the spilled material to some degree.

The following guidelines should be followed in the event of a chemical spill:

- Before any action is taken, the proper safety apparel must be worn to guard against injury. This apparel includes standard personal safety equipment as well as respirators if the spill is outside of the ventilation hood and poses a potential hazard through inhalation.
- Contain liquid spills immediately using an absorbent material (commercial spill kits, vermiculite, or sorbent pads, for example). Place the absorbent material into an appropriate container, label it clearly, and dispose of it properly.
- Use neutralizing solutions to clean up acid or base spills. After containing the spill, check the pH of the material to ensure complete neutralization.
- Use mercury sponges (commercially available) to clean up mercury spills. Following use, store the sponge in an airtight container until disposal.
- Carefully clean any equipment or containers that have been splashed to prevent future damage or injury.

GAS CYLINDERS. Gas cylinders can create hazards in the laboratory if not properly handled. All cylinders used in a laboratory must be tightly secured so that it is impossible to accidentally tip or knock them over. Regulators in use should be appropriate for the nature and pressure of the gas regulated. Each time a tank is changed, the connection between the regulator and the cylinder must be "snooped" with soapy water to detect leaks. For especially hazardous gases, detectors can be purchased that will check for any unintended release. Tanks not "in use," that is, not connected, should be identified as empty or full. Flammable gases having the potential to flash back during operation should be fitted with arrestors. Specific safety information regarding gas cylinders, including the MSDS and pressure-regulation requirements, should be sought from the supplier of the gas.

WASTE DISPOSAL. Laboratory personnel must be aware that the disposal of chemicals and other laboratory wastes may be regulated by local, state, and federal agencies. The regulations are varied, comprehensive, and often confusing for laboratory personnel who do not have specific training in the waste disposal area. Before anything is disposed of via the sewer or in containers, local agencies should be consulted. In some cases, hiring a professional to evaluate the waste generated in the laboratory and proper disposal and treatment options is warranted. A comprehensive plan needs to be established, with all laboratory workers trained in its use.

MAINTENANCE. Proper maintenance of all instrumentation and equipment must be performed and documented. The technician who uses a piece of equipment the most should be responsible for its condition. If several people share an instrument, someone must be assigned to maintaining its safe operating condition. An annual electrician's audit is useful for discovering potential problems such as frayed cords, corroded electronics, and overloaded circuits.

*F*IELD SAMPLING

TRAINING. Employees assigned to collect samples must receive the required training so that they can safely perform their tasks. The type of training required obviously depends on the type of work tasks that are assigned to the employee. The Occupational Safety and Health Administration requires specific training for employees exposed to hazardous chemicals and blood-borne pathogens. Employees who are required to use personal protective equipment such as respirators, gloves, boots, or hard hats must also be trained in the proper use and limitations of this equipment. In addition, employees working at hazardous waste sites or in confined spaces (such as manholes) must also receive specific training. The Department of Transportation regulations also require that training be given to employees who handle, package, or transport hazardous materials (49 CFR 171–180). Other training that should be given to employees includes injury prevention training such as instruction in the proper lifting of heavy items (samplers occasionally must lift heavy items such as coolers).

Occupational Safety and Health Administration regulations that may be applicable to sampling tasks include, but are not limited to, the following regulations:

- Hazardous Communication Standard—29 CFR 1910.1200 (1982),
- Bloodborne Pathogen Standard—29 CFR 1910.1030 (1991),
- Personnel Protective Equipment—29 CFR 1910.132–138 (1994),
- Hazardous Waste Operations and Emergency Response—29 CFR 1910.120 (1990),
- Permit-Required Confined Spaces—29 CFR 1910.146 (1994), and

---

- Department of Transportation Hazardous Materials Transportation Regulations—49 CFR 171–180.

MEDICAL MONITORING. Depending on the types of field tasks to which the samplers are assigned, workers may be subject to medical monitoring requirements. Occupational Safety and Health Administration regulations require medical monitoring for employees who work on hazardous waste operations sites and for those employees who wear respirators.

FIELD SAFETY. As discussed earlier, safety procedures are mainly commonsense guidelines. Many of the laboratory safety procedures are applicable to field work. The following practices must be followed by personnel conducting sampling tasks:

- Smoking, eating, chewing gum or tobacco, and drinking are forbidden except in clean or designated areas.
- Ignition of flammable liquids within or through improvised heating devices (for example, barrels) is forbidden.
- Contact with samples, excavated materials, or other contaminated materials must be minimized.
- Use of contact lenses is prohibited at all times.
- Do not kneel on the ground when collecting samples.
- All electrical equipment used in outside locations, wet areas, or near water must be plugged into outlets protected by ground fault circuit interrupters.
- A "buddy system" in which another worker is close enough to render immediate aid will be in effect.
- Good housekeeping practices are to be maintained.
- Where the eyes or body may be exposed to corrosive materials, suitable facilities for quick drenching or flushing shall be available for immediate use.
- In the event of treacherous weather-related working conditions (such as thunderstorms, limited visibility, or extreme cold or heat), field tasks will be suspended until conditions improve or appropriate protection from the elements is provided.

A hazard evaluation will be conducted before performing any field work. Site-specific safe work practices will be discussed with the sampling crew before performing any work tasks.

ACCIDENTS AND INCIDENTS

The laboratory must have a standard format for reporting all accidents and incidents. Incidents should be categorized as near-accidents—warning signs that something must be corrected in a timely fashion to avoid personal or property damage. Examples of incidents are blocked exits and/or inaccessible safety equipment, potential tripping hazards, and unmarked wet floors. All accidents must be promptly reported, regardless of the magnitude of personal injury or property loss.

The reporting forms should address three areas: the description of the incident or accident, any personal injury or property loss, and the evaluation of the occurrence by the safety manager. The description of the incident must provide the reviewer with all pertinent information. It should include the date and time, the personnel involved, the location of the incident, the work function being performed, and the nature of the incident. The safety equipment in place at the time of the incident should be listed, and an assessment of any injury or property damage should be made. The time at which the supervisor was notified must also be noted.

The injury report should be completed by affected personnel when possible. It should include the name and work identification of the injured employee. The nature of the injury must be clearly explained, including all body parts affected. Any care provided on site must be identified. If professional medical care was necessary, the name of the doctor and hospital must be provided. An example incident report is presented as Figure 9.1.

The evaluation of the incident must be thorough and should be performed by the safety director with the assistance of the personnel involved. The cause of the incident must be identified, and a decision must be made regarding its prevention. If the incident was preventable under the current safety program, mistakes or errors made must be addressed. If the incident was not preventable, then the safety program must be amended to ensure that there is no recurrence of the factors contributing to the problem. In either case, a corrective action must be developed, documented, and signed by the safety director and affected laboratory workers. The corrective action should be discussed at a safety meeting so that everyone is aware of the problem and its elimination. The responsibility for instituting the corrective action and enforcing it in the laboratory is primarily the safety director's.

It must be laboratory policy that all accidents be reported as soon as possible after the event. Laboratory workers should not judge whether an accident was too small or insignificant to be reported. Any accident and subsequent report can be useful in learning how to prevent injuries in the laboratory or in the field. It is also likely that an accident may have been preceded by a near-accident at some time. An atmosphere must be created in which the reporting of accidents is complete and prompt.

1. Personnel involved
   _____

2. Date and time of accident/incident
   _____

3. Nature of injury (including affected body part)
   _____

4. Description of property damage
   _____

5. Location of accident or incident (give address of laboratory or address of where field work assignment was being
   performed) _____

6. Describe all events leading up to accident/incident (including function being performed and injured's actions)
   _____
   _____
   _____

7. Describe the cause of the accident/incident and list any other contributing factors
   _____
   _____
   _____

8. List safety equipment in place at the time of the incident/accident
   _____

9. Was personal protection equipment required?   Was it being worn?
   _____

10. Date and time supervisor was notified
    _____

11. Date and time of treatment
    _____

12. Name and address of treating physician or attendant
    _____

13. Name and address of hospital (if applicable)
    _____

14. Describe action(s) taken to prevent a recurrence of the injury
    _____
    _____
    _____

15. Name(s) of person(s) responsible for corrective action and date actions are to be completed
    _____
    _____

16. Personnel involved in accident/incident (signature)
    _____ Date _____
    _____ Date _____
    _____ Date _____

17. Supervisor's signature
    _____ Date _____

18. Safety Director's signature
    _____ Date _____

19. Personnel responsible for corrective action signature
    _____ Date _____

Figure 9.1   Laboratory and field accident/incident reporting form.

*E*MERGENCIES

All personnel must be properly trained on what to do in cases of emergencies. In the event of a fire or severe safety hazard, workers must know the proper way to evacuate the building or job site. This requires that the location of all emergency exits or escape routes be known and that passages through them be unobstructed. If the laboratory is in an area that receives heavy snowfall, someone should be assigned the responsibility of clearing snow so that each emergency door can open freely. During times when the safety director is not in the laboratory, another person must be delegated the responsibility for handling emergency situations.

Practice drills should be held to simulate emergency evacuation. These drills will help familiarize laboratory and field workers with the sound of the evacuation alarm. An outside location where all evacuated employees are to meet should be designated. This location should be chosen so that it does not impede any necessary firefighting or rescue activities. When possible, it should be opposite in direction to the prevailing winds.

Employees should not be allowed to perform laboratory or field work without at least one other person present in the area. This backup is available to help if trouble develops. Someone should be available to make a phone call in an emergency or assist in removing injured personnel from hazards. Many local Red Cross organizations teach standard first-aid and adult cardiopulmonary resuscitation courses. Shift supervisors should be encouraged to attend these programs.

Index

A

Accidents, 185
  reporting forms, 186
Accuracy, 135
  blind standards, 72, 75
  control charts, 140
  field spike samples, 72, 75
  measurement, 136
  percent deviation, 137
  random errors, 136
  spiked samples, 137
  systematic errors, 136
Accuracy and precision sampling, 72
Activated-sludge plants
  NPDES permit discharge criteria, 23
  process testing, 23
Aerobic digestion
  sample type, 20
  sampling frequency, 20
  sampling location, 20
Anaerobic digestion
  sample type, 20
  sampling frequency, 20
  sampling location, 20
Audits, 115
Automatic sampler types, 39
  automatic samplers for VOCs, 40
  flow-through samplers, 40
  portable equipment, 39
  pressurized line samplers, 40
  refrigerated equipment, 39
  special equipment, 39
Automatic samplers, 33
  evaluation, 34
  installation and use, 37
  intake device, 34
  power and controls, 36
  pump system, 35
  sample storage requirements, 36
  sample transport, 34
  winter operation, 38

B

Bench sheet design, 148, 150–152
Billing reports, 157
Blanks, 68
  decon blanks, 70
  equipment blanks, 70, 74
  field blanks, 69, 74
  field rinseate blanks, 70, 74
  matched-matrix field blanks, 70, 74
  preparation blanks, 71, 75
  preservative blanks, 71, 75
  VOC trip blanks, 69, 74
Blind standards, 72, 75

C

Calibration, 76, 77
  blanks, 76, 77
  check standards, 76, 77
  quality assurance procedures, 78
  standards, 76, 77
Chain-of-custody forms, 107, 110

Chemical changes, 89, 90
   biological, 90
   chemical, 90
   physical, 90
Chemical spills, 182
Chlorine residual measurements, 85
Closed-pipe flow measurement, 54
   flow tubes, 56
   magnetic flow meter, 55
   orifice and segmented meters, 59
   propeller meter, 55
   rotameter, 59
   ultrasonic flow meter, 58
   variable-area meter, 58
   Venturi tubes, 57
Collection systems, 15
Collection systems studies, 5
Composite samples, 8
   fixed-volume, 8
   flow proportional, 9
   variable-frequency technique, 12
   variable-volume technique, 9, 10
Computer software, 144
Conductivity meters, 83
Contamination
   chemical changes, 89, 90
   reagents, 100
   samples contamination, 67
Cost-analysis reports, 167

D

Data
   average deviation, 142
   outlying data, 141
   Q test, 142
   standard deviation, 142
Data collection and recording, 105
   chart recording, 112
   labeling samples, 107
   personnel sampling instructions, 106
   program set-up, 106

Data handling, 146
   bench sheet design, 148
   daily summaries, 149, 153
   workbooks, 149
Data preparation, 117
Data reduction, 120
   percentile, 125
   range, 125
   rounding off, 122
   significant figures, 120
Data reporting, 145
Data reporting management
   systems, 169
Density measurement, 60
   gamma radiation, 60
   ultrasonic sensor, 60
Dewatering and thickening
   sampling, 21
   frequency, 21
   location, 21
   type, 21
Discrete sample, 7
Dissolved oxygen meters, 82
Documentation and recordkeeping, 86
Duplicates, 73, 76

E

Emergencies, 186
Equipment, 29
   automatic samplers, 33
   chain-of-custody forms, 107
   log books, 107
   manual samplers, 31
   measurement interferences, 78
   safety, 107
   sampling, 62, 98
Error
   analytical interferences, 66
   sample contamination, 67

L

Labeling samples, 107
Laboratory identification, 109
Laboratory information
    management system (LIMS),
    169
Laboratory reports, 155
Laboratory work procedures
   safety, 179
      chemical spills, 182
      chemicals, 180
      food and beverages, 179
      fume hoods, 181
      gas cylinders, 182
      glassware, 179
      maintenance, 183
      waste disposal, 183
Limit of detection value, 142
Linear regression, 140

M

Manual samplers, 31
   Coliwasa sampler, 32
   pond sampler, 32
   weighted bottle sampler, 32
Meters
   area-velocity meter, 54
   conductivity, 83
   dissolved oxygen, 82
   magnetic flow meter, 55
   orifice and segmented , 59
   pH, 81
   propeller meter, 55
   rotameter, 59
   ultrasonic flow, 58
   variable-area, 58

N

National Pollutant Discharge
    Elimination System (NPDES),
    4, 9
   reports, 160
Noncompliance reports, 167
Nondetect results, 143
Nonliquid sampling, 60
   gases, 60
   shunt meter, 60

O

Occupational Safety and Health
    Administration, 183
Open-flow nozzles, 46

P

Personnel
   data collection and recording
      instructions, 106
   medical monitoring, 184
   safety, 5
   sample collection training, 30
pH meters, 81
Planning, 6, 91
   containers, 93
   holding times, 93
   preservation techniques, 93
   sample equipment, 98
   sample handling, 98
Plant operations reports, 157
Precision, 72, 136
   average deviation, 138
   control charts, 140
   duplicates, 73, 76
   field split samples, 73, 76
   measurement, 137
   standard deviation, 138
   variance, 138